SpringerBriefs on Cyber Security Systems and Networks

The series aims to develop and disseminate an understanding of innovations, paradigms, techniques, and technologies in the contexts of cyber security systems and networks related research and studies. It publishes thorough and cohesive overviews of state-of-the-art topics in cyber security, as well as sophisticated techniques, original research presentations and in-depth case studies in cyber systems and networks. The series also provides a single point of coverage of advanced and timely emerging topics as well as a forum for core concepts that may not have reached a level of maturity to warrant a comprehensive textbook. It addresses security, privacy, availability, and dependability issues for cyber systems and networks, and welcomes emerging technologies, such as artificial intelligence, cloud computing, cyber physical systems, and big data analytics related to cyber security research. The mainly focuses on the following research topics:

Fundamentals and Theories

- Cryptography for cyber security
- Theories of cyber security
- Provable security

Cyber Systems and Networks

- Cyber systems security
- Network security
- Security services
- Social networks security and privacy
- Cyber attacks and defense
- Data-driven cyber security
- Trusted computing and systems

Applications and Others

- Hardware and device security
- Cyber application security
- Human and social aspects of cyber security

More information about this series at http://www.springer.com/series/15797

Shankar Karuppayah

Advanced Monitoring in P2P Botnets

A Dual Perspective

Shankar Karuppayah
National Advanced IPv6 Centre (NAv6)
Universiti Sains Malaysia
USM, Penang
Malaysia

ISSN 2522-5561   ISSN 2522-557X (electronic)
SpringerBriefs on Cyber Security Systems and Networks
ISBN 978-981-10-9049-3   ISBN 978-981-10-9050-9   (eBook)
https://doi.org/10.1007/978-981-10-9050-9

Library of Congress Control Number: 2018940630

Printed on acid-free paper

This Springer imprint is published by the registered company Springer Nature Singapore Pte Ltd.
The registered company address is: 152 Beach Road, #21-01/04 Gateway East, Singapore 189721, Singapore

எல்லாம் புகழும் இறைவனுக்கே!

All Praise To God!

Foreword

Botnets—is it exaggerated to call them one of the most dangerous weapons of mass destruction? They represent definitely the most low-cost one, and not general nor soldier has to get his hands dirty using them: a little silent and unnoticed invasion in your and my computer, smartphone, or smart TV, and whoops!—your hardware has become part of a gigantic evil army. Maybe while you read this, your own hardware is armed for a million- or billion-computer attack, helping to find entry holes into critical servers, to turn the computers of even more people into slave soldiers, to paralyze the entire infrastructure of an enterprise, or to confuse the metro system of a big city. Today, most attacks still only target the IT infrastructure of big service providers and enterprises, health systems, governments, etc., yet we witness already increasingly dramatic effects: National elections become suspicious due to successful attacks, ransoms are demanded after virtually disabling big health infrastructures by blocking their administrative IT, huge financial losses are triggered by paralyzing fully IT-dependent services like banks or online shops, and government secrets are digitally spied by potential enemies.

Ironically, the ever-continuing digitization of our society is currently readying even more crucial sectors of our life for those new weapons of mass destruction. One aspect of this is critical infrastructures that will soon be fully dependent on networked IT: For instance, the complex decentralized energy production and storage that marks the smart grid evolution will hardly be controllable without interconnected IT acting as the nervous system of this complex body. What is more, the Internet of things and the ongoing fourth wave of the industrial revolution will potentially bring forth a direct attack path for botnets to every facet of our economy, even our very lives.

In light of these huge global threads, works like the present monograph written by Shankar Karuppayah are desperately needed. Why? Because it is great time to form much more forceful armies of defense against the huge evil botnet armies that threaten us. Some of these armies shall be at the same scale as botnets, acting as globally operating defense armies. Some of them will be small but with highly effective stealthy special missions, capable of shutting down an evil army with a sneak attack. Like the botnets themselves, these armies will mainly consist of hard- and software, and their success will depend on the capabilities of the brains that design and launch them. As we need many more highly capable researchers and developers on the defense side, we have to have accurate and detailed means of knowledge transfer that put these brains in a state of superior skill and understanding effectively and efficiently.

As a first and crucial step toward this aim, the first part of Mr. Karuppayah's book provides general background knowledge, terminology, and insights into related work in his first two introductory chapters. His third chapter delivers extremely illustrative and enlightening insights into three very relevant sample botnets. The following chapters four and five represent the very core of the effort that this book represents: Here, Mr. Karuppayah treats the two crucial measures for infiltrating the evil armies, as a prerequisite for any effective countermeasure. The first approach is called crawling—it tries to elicit information by talking to botnet nodes in their own jargon. With the second approach, called sensing, defenders become integral parts of the attack armies, i.e., botnets. In both cases, Mr. Karuppayah also takes on the hat of the dark side, trying to determine how the attackers would react to his infiltration measures. Then, in turn, he takes the perspective of the defense army again, trying to retain an edge in the arms race. This changing of perspectives is an indispensable quality of any brain working on botnet defense, a quality that can best be learned by example, and Mr. Karuppayah's example is a model one to learn from.

In conclusion, one can consider the present book as small but extremely valuable prerequisite for anyone who wants to defend our civilization on maybe the most decisive and critical battlefields of our time a battlefield that is becoming the central arena for war, espionage, terrorism, and crime alike. In short, one can truly call it a timely book on one of the biggest evils of our time.

Darmstadt, Germany   Prof. Dr. Max Mühlhäuser
March 2018   Full Professor (Computer Science)
                                           Head of Telecooperation Lab, TU Darmstadt

Preface

Botnets are increasingly being held responsible for most of the cybercrimes that occur nowadays. They are used to carry out malicious activities like banking credential theft and DDoS attacks to generate profit for their owner, the botmaster. Recent botnets have been observed to prefer P2P-based architectures to overcome some of the drawbacks of the earlier architectures.

The distributed nature of such botnets requires the defenders, i.e., researchers and law enforcement agencies, to use specialized tools such as crawlers and sensor nodes to monitor them. However, realizing this, botmasters have introduced various countermeasures to impede botnet monitoring, e.g., automated blacklisting mechanisms. The presence of anti-monitoring mechanisms not only renders monitoring data to be inaccurate or incomplete, but may also adversely affect the success rate of botnet takedown attempts that rely upon such data. Most of the existing monitoring mechanisms identified from the related works only attempt to *tolerate* anti-monitoring mechanisms as much as possible, e.g., crawling bots with lower frequency. However, this may also introduce noise into the gathered data, e.g., due to the longer delay in crawling the botnet. This in turn may also deteriorate the quality of the data.

This book addresses most of the major issues associated with monitoring in P2P botnets as described above. Specifically, it analyzes the anti-monitoring mechanisms of three existing P2P botnets—(1) *GameOver Zeus*, (2) *Sality*, and (3) *ZeroAccess*—and proposes countermeasures to circumvent some of them. In addition, this book also proposes several advanced anti-monitoring mechanisms from the perspective of a botmaster to anticipate future advancement of the botnets. This includes a set of lightweight crawler detection mechanisms as well as a set of novel mechanisms to detect sensor nodes deployed in P2P botnets. To ensure that the defenders do not loose this arms race, this book also includes countermeasures to circumvent the proposed anti-monitoring mechanisms.

The works discussed in this book have been evaluated using either real-world botnet datasets; i.e., those were gathered using crawlers and sensor nodes, or simulated datasets. Evaluation results indicate that most of the anti-monitoring mechanisms implemented by existing botnets can either be circumvented or tolerated to obtain better quality of monitoring data. However, many crawlers and sensor nodes in existing botnets are found vulnerable to the anti-monitoring mechanisms that are proposed from the perspective of a botmaster in this book. Existing and future botnet monitoring mechanisms should apply the findings of this book to obtain high-quality monitoring data and to remain stealthy from the bots or the botmasters.

Penang, Malaysia   Shankar Karuppayah
March 2018

Acknowledgements

This book would not have come into existence without the help and encouragement of my colleagues, friends, and family. My heartfelt gratitude and thanks go out to all of them.

Firstly, I would like to express my sincere gratitude to my Doktorvater Max Mühlhäuser for the continuous support during my Ph.D. studies, his patience, motivation, and immense knowledge. His guidance helped me in all the time of research and completing my Ph.D. I could not have imagined having a better advisor and mentor for my Ph.D. study. Secondly, I am indebted to the supervision of Mathias Fischer since the beginning. The various discussions, experiences, and guidance I received from him are definitely unforgettable and in part helped to shape this book. I am also thankful to the various co-researchers that I have had the privilege to collaborate in some of the works presented in this book.

I would also like to thank all of my current and former colleagues and students at the National Advanced IPv6 Centre of Universiti Sains Malaysia and at the Telecooperation Group of TU Darmstadt, who have contributed directly or indirectly toward the completion of this book. Last but not least, I would like to thank my parents Mr. C. Karuppayah and Mrs. K. Anjama, my wife Prevathe Poniah, my brothers, and sisters-in-law for being my support system.

Without all of your support, this book would not have been successful.

Contents

Acronyms

BFS	Breadth-first search
BT	Booby trap
C2	Command-and-control server
DDoS	Distributed denial of service
DFS	Depth-first search
DGA	Domain generation algorithm
DHCP	Dynamic Host Configuration Protocol
DHT	Distributed hash table
DNS	Domain Name System
DSI	Distributed sensor injection
HTTP	Hypertext Transfer Protocol
HTTPS	Hypertext Transfer Protocol Secure
IDS	Intrusion detection system
IP	Internet Protocol
IPS	Intrusion prevention system
IRC	Internet Relay Chat
ISP	Internet service provider
LCC	Local clustering coefficient
LICA	Less Invasive Crawling Algorithm
MM	Membership maintenance
NAT	Network address translation
NL	Neighbor list
OSN	Online social network
P2P	Peer-to-peer
SCC	Strongly connected component
UDP	User Datagram Protocol
URL	Uniform Resource Locator

Chapter 1
Introduction

Banking fraud, spam campaigns, and denial-of-service attacks are types of cyber-crimes that are profitable business. These attacks are often carried out using botnets, a collection of vulnerable machines infected with malware that are controlled by a botmaster via a Command-and-Control Server (C2). Traditional botnets utilize a centralized architecture for the communication with the botmaster. However, if such a (C2) is taken down, the botmaster cannot communicate with its bots anymore. Recent Peer-to-Peer (P2P)-based botnets, e.g., *GameOver Zeus* [1], *Sality* [2], or *ZeroAccess* [3], adopt a distributed architecture and establish a communication over-lay between participating bots. Due to the lack of central entities, such botnets are much more resilient to attacks than centralized botnets. In fact, all (counter-)attacks against P2P-based botnets require detailed insights into the nature of these botnets, in particular the botnet population and the connectivity structure among the bots [4]. As a consequence, monitoring such botnets is an important task for analysts.

Since botnets are valuable assets to the botmasters, they often attempt to impede the performance of monitoring mechanisms. Although some of the existing and proposed mechanisms are still in their infancy, it is just a matter of time before more advanced countermeasures are introduced.

In the following, some essential background information that is useful to under-stand the rest of the chapters is presented.

1.1 Botnet Architectures

A botnet consists of infected machines or bots that are controlled by a botmaster via a *Command-and-Control Server (C2)*. The C2 is used to disseminate new configura-tions and updates to the bots and also to upload stolen data from the infected machines, e.g., credit card credentials or passwords. A botnet can be classified according to three architectures as described in the following.

S. Karuppayah, *Advanced Monitoring in P2P Botnets*, SpringerBriefs on Cyber Security Systems and Networks, https://doi.org/10.1007/978-981-10-9050-9_1

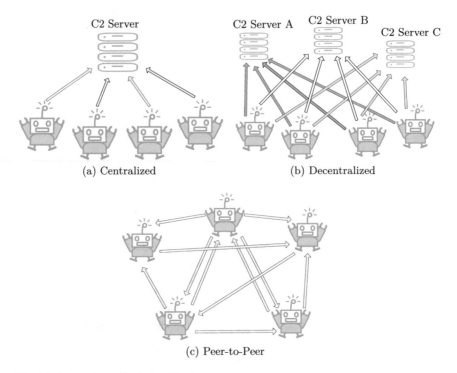

Fig. 1.1 Comparison of botnet architectures

1.1.1 Centralized Botnets

Traditional botnets were observed to utilize centralized C2s. The C2s are often deployed either on self-deployed Internet Relay Chat (IRC) servers or hacked web-servers. The network topology of such botnets are depicted in Fig. 1.1a. Bots in centralized botnets regularly poll the C2s for newer updates from botmasters and apply them as soon as they are available. Although centralized C2s are easy to deploy and have low communication latency, the architecture poses itself as a single point of failure, e.g., botnet takedown attempts. The removal of the C2 server renders the entire botnet incapable of retrieving commands or contacting the botmaster. Moreover, defenders can also easily enumerate all infected machines using only the server's communication logs.

1.1.2 Decentralized Botnets

The weakness of centralized botnets led to the next generation of botnets to adopt a decentralized architecture where the simplicity and efficiency of a centralized architecture is retained as much as possible but with improved resilience against botnet takedowns by adding redundant C2s as depicted in Fig. 1.1b. Botmasters were observed to experiment with several strategies to implement the redundant C2 feature. Most commonly, the bots have several hard-coded C2 address that are contacted sequentially if an existing C2 is not reachable.

More advanced strategies utilized the implementation of Domain Generation Algorithm (DGA) or fast-fluxing mechanisms. A DGA is an algorithm that generates time-sensitive domain names of C2s using a common seed, e.g., date of the day, across the different bots. This way, the botmaster is able to register many domain names in advance that would be contacted by the bots in the future. Even if some of the C2s are taken down, the botmaster is still able to re-establish communication via future domain names. However, since the DGA is often hard-coded in the bot's binary, through reverse-engineering, one can identify future domain names that will be used in the bots. Therefore, defenders are able to find out and register these domain names to hijack the botnet from the botmasters [5].

Another variant of the redundancy strategy in C2s is implemented using Domain Name System (DNS) fast-fluxing networks. In such networks, the IP addresses of multiple C2s is cycled rapidly via the usage of DNS records for a domain name. In contrast to DGA-based mechanisms, if the botmasters are in control of the fluctuation of the IPs, defenders are not able to predict which IP address will be used to point bots to a C2 in the future. Moreover, an advanced variant of fast-fluxing is called double fast-fluxing: it cycles a list of name servers which are utilized by the bots to resolve the C2's domain name. Such a design introduces an additional layer of fluctuation to increase resiliency and to prevent botnet takedowns.

1.1.3 P2P Botnets

Recent botnets are observed to adopt a P2P-based architecture which eliminates vulnerabilities that were present in the other architectures. Bots in P2P botnets are interconnected via an *overlay* that consists of neighborhood relationships between a bot and a subset of other bots as depicted in Fig. 1.1c. This overlay is maintained using a *Membership Maintenance (MM)* mechanism that is botnet-specific (cf. Sect. 3). Like traditional P2P networks, P2P-based botnets also experience *node churn*, i.e., nodes joining and leaving the network at high frequency. To withstand churn, the *MM* mechanism ensures that participating bots remain connected to the overlay by ensuring that unresponsive bots, e.g., offline bots, are removed from the *Neighborlist (NL)* of the bot and replaced by responsive ones.

In a P2P botnet, a botmaster can pick any bot within the botnet to *inject* commands which are eventually disseminated to all bots, albeit with a higher delay. Nevertheless, the ability to inject commands to any bot in the network is advantageous not only because it allows the botmaster to have many entry points but it also cloaks the source of the command to prevent any traceback attempts.

Although P2P botnets are very resilient, takedown attempts on P2P botnets are still possible through sinkholing attacks that require a vulnerability within the botnet's design or communication protocols. However, such an attack requires enumeration information of all bots in the botnet using monitoring mechanisms as presented in the following section.

1.2 P2P Botnet Monitoring

The most common P2P botnet monitoring mechanisms are the *honeypot*, *crawlers*, and *sensors*. Honeypots are special systems with the only purpose of being infected by a malware so that subsequent communication of the honeypot can be logged for analysis (cf. Sect. 2.3.1). Although honeypots are easy to deploy, they can only gather limited information about the bots in a botnet.

In contrast, crawlers leverage the botnet's communication protocol to iteratively request neighbors of bots until all bots have been discovered (cf. Sect. 2.3.2). However, this monitoring mechanism requires a complete reverse engineering of the botnet's binary to understand and re-implement the botnet's communication protocols to send and receive valid botnet-messages. The major drawback of crawlers is the inability to contact bots behind network devices that implement Network Address Translation (NAT) [6]; such bots which are not directly reachable from the Internet, i.e., non-routable, represent the majority segment of a botnet's population, i.e., up to 90% [4].

Sensors are deployed to address the drawbacks of crawlers (cf. Sect. 2.3.3). Sensors are deployed using public IP addresses to enable all bots to contact them directly. Since non-routable nodes in P2P networks remain connected to the botnet overlay through nodes with public IP addresses, sensors will eventually be contacted by the non-routable bots. Therefore, sensors aim to be responsive to all bots contacting them to remain in the NL of bots as reliable neighbors, i.e., always online and responsive. By responding to incoming request messages, sensors can enumerate both routable and non-routable nodes based on their identities, e.g., IP address and/or botnet specific identifiers. However, a major drawback with sensors is that they often cannot obtain the inter-connectivity among bots.

1.3 Outline

The remainder of this book is outlined as follows: Chap. 2 presents the requirements for any botnet monitoring mechanism and a formal model for P2P botnets. In addition, the chapter also discusses the state of the art in both botnet monitoring and anti-monitoring mechanisms. Chapter 3 presents the summary analysis of reverse engineering of three major P2P botnets that are focused in this book.

The major contributions of this book are presented in Chaps. 4 and 5. Chapter 4 presents works that improve efficiency of existing crawlers as well as proposals that botmasters could use in the future to impede crawler capabilities. Similarly, Chap. 5 introduces novel sensor detection mechanisms to discern sensors from bots and also proposals to circumvent them as well. Finally, Chap. 6 presents a summary of this book and details the future work in the field of advanced P2P botnet monitoring.

References

1. Andriesse, D., Rossow, C., Stone-Gross, B., Plohmann, D., Bos, H.: Highly resilient Peer-to-Peer botnets are here: an analysis of Gameover Zeus. In: International Conference on Malicious and Unwanted Software: The Americas (2013)
2. Falliere, N.: Sality: Story of a Peer-to-Peer Viral Network. Technical report, Symantec (2011)
3. Wyke, J.: The ZeroAccess BotnetMining and Fraud for Massive Financial Gain. Sophos Technical Paper (2012)
4. Rossow, C., Andriesse, D., Werner, T., Stone-gross, B., Plohmann, D., Dietrich, C.J., Bos, H., Secureworks, D.: P2PWNED: modeling and evaluating the resilience of Peer-to-Peer botnets. In: IEEE Symposium on Security and Privacy (2013)
5. Stone-gross, B., Cova, M., Cavallaro, L., Gilbert, B., Szydlowski, M., Kemmerer, R., Kruegel, C., Vigna, G.: Your botnet is my botnet : analysis of a botnet takeover. In: ACM CCS. ACM (2009)
6. Egevang, K., Francis, P.: The IP network address translator (NAT). Technical report, RFC 1631 (1994)

Chapter 2
Requirements and State of the Art

This present chapter provides a discussion on botnet monitoring as well as the common challenges. In more detail, Sect. 2.1 presents the requirements of a botnet monitoring mechanism with an emphasis on P2P botnets. Then, Sect. 2.2 presents a formal model on P2P botnets that is used throughout this book.

In Sect. 2.3, the related work in botnet monitoring is discussed. After that, Sect. 2.4 discusses the challenges commonly faced in botnet monitoring. In particular, Sect. 2.4.1 introduces issues stemming from the dynamic nature of P2P networks. Section 2.4.2 elaborates on the pollution of monitoring results due to monitoring activities of unknown third parties. In addition, Sect. 2.4.3 discusses the various anti-monitoring mechanisms implemented in botnets and those proposed by related work. Finally, a thorough discussion sums up the chapter in Sect. 2.5.

2.1 Requirements of a Botnet Monitoring Mechanism

In the following, the functional and non-functional requirements of a botnet monitoring mechanism are presented.

2.1.1 Functional Requirements

Any botnet monitoring mechanism has to conform to the following functional requirements:

1. **Genericity**: To ensure adaptability across different botnets, a monitoring mechanism should be designed and developed in a generic manner. Hence, the

© The Author(s), under exclusive license to Springer Nature Singapore Pte Ltd., part of Springer Nature 2018
S. Karuppayah, *Advanced Monitoring in P2P Botnets*, SpringerBriefs on Cyber Security Systems and Networks, https://doi.org/10.1007/978-981-10-9050-9_2

monitoring mechanism should be easily adapted to any P2P botnets by only implementing the necessary communication protocols.

2. **Protocol Compliance**: Monitoring mechanisms should comply with the protocols of the botnet under scrutiny. Compliance is important as most botnets respond only to valid messages that strictly adhere to the protocols, e.g., encryption and decryption routines.

3. **Enumeration of Bots**: Enumeration capability is an important aspect of a monitoring mechanism. Through enumeration, it is possible to estimate the population size of the botnet. Bot enumeration is usually done by leveraging on the botnet-specific request and reply messages. Every valid response or an unsolicited request message from a bot indicates the presence of an active bot.

4. **Neutrality**: A monitoring mechanism should stay neutral during monitoring to avoid introducing artificial noise that may taint the observed behavior of a botnet. Specifically, a mechanism should avoid executing command from the botmaster or disseminating them further to other bots.

   Besides that, the mechanism should also avoid introducing noise that may disrupt or hamper the normal activities of the bots. Disrupting the activity of the bots will introduce bias or noise in the monitoring data and may lead to inaccurate conclusion of the real nature of the botnet.

5. **Logging**: All information gathered from a monitoring mechanism should be logged alongside the associated timestamps. The log should include additional botnet-specific metadata such as the details of the latest command known to the bot or its unique identifiers (if available).

   The logged information is particularly useful to inform the relevant stakeholders, i.e., Internet Service Providers (ISPs) and network administrators, about the infections as well as to understand the botnet itself. For instance, the information from a crawler can be used to reconstruct the botnet's network topology from the crawler's point of view. This information can then be further analyzed using graph analysis techniques to identify most influential bots in the botnet overlay.

2.1.2 Non-functional Requirements

The following non-functional requirements are directly related to the quality of a botnet monitoring mechanism as well as its collected data.

1. **Scalability**: A monitoring mechanism may be defined as scalable when its performance does not deteriorate with an increased number of bots in a botnet. In existing and earlier botnets, the total population ranged anywhere between several thousand to a few million bots. However, scalability of a monitoring mechanism is not only about handling high volume of request and reply messages of bots in a botnet, but also system resources regarding memory, bandwidth, computational resources and storage space.

2. **Stealthiness**: It is important to ensure that a mechanism is not identifiable during monitoring. Since monitoring activities threaten the economy of the botnets themselves, botmasters may retaliate against such monitoring mechanisms, e.g., DDoS attack. The retaliation attacks may cause disruption to an ongoing monitoring activity or renders the IP address completely unusable for further monitoring.

3. **Efficiency**: Efficiency is two-fold and can be observed as *probing* and *resource efficiency*. For probing activities, efficiency manifests on the ability of minimizing noise introduced in the resulting monitoring data, e.g., a delay in probing or crawling bots may lead to bias in the resulting data [26].

   The resource efficiency focuses on the ability to perform monitoring with minimal resources, e.g., minimum number of request messages. A crawler could omit sending the (optional) probe messages and utilize the neighborlist request/reply message instead to assert the responsiveness of a bot as well as to obtain the neighbors of the bot.

4. **Accuracy**: Accuracy is two-fold and can be observed in *enumeration* and *connectivity* accuracy. For bots enumeration, accuracy manifests in the ability to discover all bots and identify those that are online or offline at a given point in time.

   Besides bot enumeration, the (inter-)connectivity among the bots is also important, e.g., for botnet takedown operations that involves strategically invalidating the inter-connectivity among bots. Hence, the connectivity accuracy is the ability to capture the exact botnet topology at a given point in time.

   Monitoring activities of others may also introduce noise in the resulting monitoring data. For instance, a sensor may yield very high *uptime* compared to regular bots and consequently affect churn measurements that are relying upon the *uptime* of bots. As such, monitoring activities should be identified and their footprint should be removed from the monitoring data.

5. **Minimal Overhead/Noise**: A monitoring mechanism should ensure that its activities or footprints do not affect the botnet by significantly alter the nature or behavior for both the botnet itself and other monitoring parties. Although it is evident that existing monitoring activities will introduce noise, it is crucial to ensure that necessary steps to reduce the noise are taken.

2.2 Formal Model for P2P Botnets

This section introduces a formal model for P2P botnets that will be useful to understand the various work presented in Chaps. 3–5.

A P2P botnet can be modeled by a directed graph $G = (V, E)$, which is a common practice [5, 6, 21], where V is the set of *bots* in the botnet and E is the set of edges or inter-connectivity between the bots, i.e., the neighborhood relationship. Bots V in a botnet can be further classified into two different categories of bots, i.e., $V = V_s \cup V_n$, : *superpeers* V_s are bots that are directly routable and *non-superpeers* V_n for those that are not directly routable, e.g., behind stateful firewalls, proxies, and network devices that use NAT. Please note that in the remainder of this book, the terms *bot*, *peer*, and *node* are used synonymously.

All bots use a MM mechanism (cf. Sect. 1.1.3) that establishes and maintains neighborhood relationships, i.e., a neighborlist, to ensure a connected botnet overlay. Hence, bots have connections to a subset of other bots, i.e., neighbors, in the overlay. These connections or edges $E \subseteq V \times V$, are represented as a set of directed edges (u, v) with $u, v \in V$. The neighborlist NL of a bot $v \in V$ is defined as $NL_v = \{u \in V | \forall u \in V : (v, u) \in E\}$. Hence, the *outdegree* of a bot v can be defined as the number of outgoing edges or neighbors maintained by the bot: $\deg^+(v) = |NL_v|$. The maximum outdegree is governed by the global botnet-specific value of maximum entries NL^{MAX} that can be stored at any given point in time, i.e., $|NL_v| \leq NL^{MAX} \leq |V|$. Moreover, the *popularity* or *indegree* of a bot v in the botnet can be measured based on the number of bots that have v as an entry in their NLs: $\deg^-(v) = |(u, v) \in E|$.

Bots in a P2P botnet use a MM mechanism to maintain their NLs regularly following a botnet-specific interval that is often referred to as MM-cycle. In each cycle, a bot probes for the *responsiveness* of all of its neighbors by sending a *probe message* to each of them. The responsiveness of neighbors can be verified based on a valid response to the sent messages. In this formal model, this probe message is referred to as the *probeMsg*.

If the bot has low amount of neighbors or many neighbors are not responsive, additional neighbors can be requested to fill up the NL. For this purpose, a peer v can request its neighbor u to select a subset of u's neighbors $L \subseteq NL_u$ and share them with v via a neighborlist request method that is referred to as the *requestL* in this model. The decision on which exact entries are picked in the returned response message L depends on the *neighbor selection criteria* employed by the botnet.

Bots (re)joining the overlay often *announce* their existence to a subset of existing superpeers using a message that is referred to as *announceMsg* in this proposed model. The information of these superpeers can originate from a hard-coded list within the malware binary for new bots or from previous bot communications for bots rejoining the overlay. Superpeers receiving such a message will check if the new bot is a potential superpeer by sending a message to the port that is used by the sender for receiving incoming requests. The information about which port to check is often transmitted along with the initial message to the superpeer. A valid response to the message indicates the new bot is a potential superpeer candidate as well. Therefore, the information of the new superpeer can be stored within the existing superpeer's NL and further propagated when being requested by other bots in need of new neighbors.

Non-superpeers would fail to receive the probe messages sent by the superpeers due to the presence of NAT-like devices that drop unsolicited messages, i.e., messages initiated remotely. Such bots are often not included in the neighborlists of the superpeers. These bots only rely upon existing superpeers to relay any update to or from the botmaster.

2.3 Related Work on Botnet Monitoring

The open nature of P2P botnets allows anyone with knowledge of the botnet communication protocols to participate and communicate with bots in the network (cf. Sect. 1.2). This openness is often exploited to monitor P2P botnets by disguising as another bot. P2P botnet monitoring is often done with the aim of identifying and enumerating all infected machines. Besides using the monitoring data to perform cleanup activities, the data can also provide valuable information in understanding P2P botnets.

Monitoring also allows understanding the modus-operandi of the botnets themselves. For instance, by analyzing the commands that are regularly issued in the botnet, it may be possible to identify the operators. Monitoring data is also useful and important in the event of a botnet takedown attempt to ensure a higher success rate [21].

In the following, the state of the art of three common botnet monitoring mechanisms: *honeypots*, *crawlers*, and *sensor nodes*, are presented with respect to the requirements presented in Sect. 2.1.

2.3.1 Honeypots

Honeypots or *honeynets* are machines or a network of machines designed to appear as lucrative targets in the eyes of malware and attackers. Such machines or networks aim at being infected to monitor any subsequent malicious activities [23]. Using honeypots is easy and straightforward as no prior knowledge of the malware or its communication protocol is required to conduct monitoring.

A malware-infected honeypot would contact its C2, e.g., IRC, to communicate with the botmaster. By monitoring the network traffic generated by the honeypot, it is possible to identify the C2, e.g., IRC server, that is being used. Moreover, the information about other infected machines that are contacting the C2 can also be obtained by inspecting the C2 communication logs [19].

The main disadvantage of honeypots is that they have minimal control over the actions that are taken by the bots within the honeypot environment, e.g., participating in an ongoing attack. In [19], the author reported his efforts to rate-limit the generated network traffic and manual blocking of certain ports to minimize the damage that may be done by the malware. These efforts are taken mainly because it may be illegal if a user knowingly *volunteers* to be part of the botnet activities or participate in an ongoing attack. Besides that, if the traffic generated by the infected honeypot is encrypted, only communication metadata can be obtained.

The limitations associated with honeypots have led to the development of more advanced monitoring mechanisms like crawlers and sensor nodes. These mechanisms allows more control over monitoring activities like selectively refuse to respond or forward certain messages, e.g., new botmaster commands. Furthermore, these mechanisms can communicate with bots and increase their monitoring coverage appropriately. Therefore, in the remainder part of this book, the focus is only on the usage of these two advanced mechanisms in the context of monitoring P2P botnets.

2.3.2 Crawlers

Due to the self-organizing nature of P2P botnets, bots can request additional neighbors when the number of responsive neighbors in their NL is low. This observation is exploited by the *crawler*, which is a computer program that mimics the behavior of a bot that is low on neighbors and request additional neighbors. Since bots only respond to communication messages that conforms to their protocol, a crawler needs to implement parts of the botnet protocol for sending valid *requestL* request messages and parsing the replies accordingly (Functional Requirement 2 and 4).

Starting with a list of *seed* nodes, i.e., superpeers, that is retrieved by reverse engineering the malware binary, a crawler requests the neighbors of this node and iteratively sends *requestL* to all newly discovered bots to obtain their respective NL as depicted in Fig. 2.1. The goal of crawling is to obtain an accurate *snapshot* of the botnet by identifying all infected bots as well as the inter-connectivity among them. Each snapshot is a directed graph that contains information of all discovered bots along with their neighborhood relationships as described in the formal model (cf. Sect. 2.2). A visualization of such a snapshot is presented in Fig. 2.2.

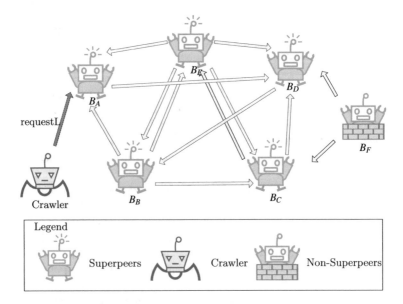

Fig. 2.1 A crawler requesting the neighborlist of bot B_A

Fig. 2.2 GameOver Zeus
network connectivity graph
among 23,196 nodes
reconstructed via crawling.
The blue dots indicate the
nodes (systems infected with
GameOver ZeuS) and the
green lines indicate the edges
between nodes (Source: Dell
SecureWorks.)

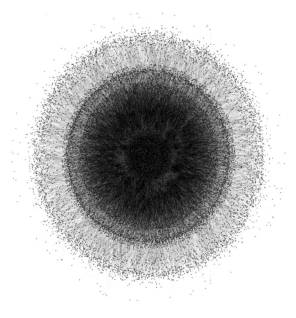

Requesting NL of all bots can be implemented using graph traversal techniques such as Depth-First Search (DFS) or Breadth-First Search (BFS). These techniques can be implemented within the crawlers by using either a stack or a queue-based implementation as the node selection strategy. Finally, the information of *who knows whom* can be stored for further analysis (Functional Requirement 5). Take note that it is important for a crawler to request the NL of bots in quick successions to reduce the network bias, e.g., address aliasing, churn effects introduced in the resulting snapshot [15, 27] (cf. Sect. 2.4.1). Accurate snapshots are important to conduct an effective botnet take-down attempt or to analyze the resilience of the botnet against attacks (Non-Functional Requirement 3 and 4).

However, crawling often fails to enumerate *all* bots in the botnet. Depending on the MM mechanism of the some botnets, crawlers can only contact the *superpeers*, i.e., bots that are directly reachable. Bots that are behind network and security devices such as NAT, proxies, and firewalls, i.e., non-superpeers, are not directly reachable by crawlers, but make up the majority of the entire botnet population (according to [21] 60–90%). As such, in our example in Fig. 2.1, the crawler would not be able to discover nor communicate with bot B_F.

One of the first P2P botnets that have caught the attention of the public and media was *Storm* [11]. This botnet initially coexisted together with the OVERNET P2P file sharing network, and eventually moved on to a bot-only network which is referred to as *Stormnet*. From then on, many researchers attempted to monitor Storm and presented their analysis [8, 11, 14, 27]. The results presented by the researchers were interesting due to discrepancies stemming from differing approaches and assumptions in monitoring the botnet [21, 31].

Since Storm initially coexisted with OVERNET, one of the main challenges was to distinguish bots from clients of OVERNET. Wang et al. presented a methodology to identify bots based on the observation that the DHT IDs of the Storm bots are not persistent over reboots and, therefore, change frequently compared to benign users [27].

One of the earliest work on monitoring Storm was performed by Holz et al. using a BFS-based crawler called *StormCrawler* [11]. This crawler iteratively queries each bot starting from a *seedlist* and sends 16 route request messages consisting of carefully selected DHT IDs, i.e., evenly spaced around the DHT space, to increase the chances of retrieving undiscovered peers. Due to the open nature of P2P botnets, many researchers started actively monitoring Storm and experimented with it. Kanich et al. reported the presence of many unknown third parties that monitor the botnet in parallel by using a built-in heuristic on their *Stormdrain* crawler to identify non-bots. This heuristic is based on a design flaw of the OVERNET ID generator within the binary that is capable of only generating a small range of IDs [14]. However, the authors admitted that they were not able to identify researchers that could have chosen IDs that fall within the range that is used by bots in Storm.

Most crawlers on Storm have been reported to conduct crawling by randomly *searching* for IDs around the DHT space in the hope of eventually discovering all participating bots. However, as reported by Salah and Strufe [22], a more accurate snapshot can be obtained using their *KAD Crawler* that crawls the *entire* KAD network in a distributed manner by leveraging the design of KAD itself, which is similar to the design adopted by OVERNET. Their distributed crawling is not dependent on *online* nodes, unlike existing Storm crawlers, instead splits the KAD ID space into multiple zones and assigning crawlers to dedicated zone to retrieve the routing tables of all nodes within each zone. The results from the different zones can then be aggregated as the snapshot of the botnet.

Other P2P botnets have also attracted the attention of researchers in monitoring them. For instance, Dittrich and Dietrich deployed a DFS-based crawler to crawl *Nugache* [7]. Their crawler conducts *pre-crawls* and utilizes that information as an input for their priority-queue based implementation that prioritizes nodes which have been observed more often available and responsive in the pre-crawls. More recently, Rossow et al. presented their analysis on the resiliency of P2P botnets, namely *GameOver Zeus*, ZeroAccess, and Sality, using their BFS-based crawler that starts crawling from a *seednode* and appends newly discovered nodes from previously crawled bots at the end of the queue. In 2014, Yan et al. introduced *SPTracker* to crawl the three botnets mentioned above [31, 32]. In contrast to conventional crawling, SPTracker includes *node injection* (cf. Sect. 2.3.3) as a complementary mechanism to obtain better crawl results.

From the domain of unstructured P2P file sharing network, Stutzbach et al. presented *Cruiser* to crawl *Gnutella* [26]. This crawler prioritizes *ultrapeers* from the two-tier design of the *Gnutella* network and can quickly capture an accurate representation of the P2P network. The authors also report of an observed connectivity bias among peers which are most likely connected to peers with higher uptime.

Similar observations in unstructured P2P botnets have also been reported by other researchers [21, 32].

Although most of the proposed botnet crawlers meet all functional requirements presented in Sect. 2.1.1, not much focus have been given on the non-functional requirements such as stealthiness, accuracy and the amount of noise being introduced within the footprint of the botnet (cf. Table 2.2). These unmet requirements will become more apparent upon discussing the challenges encountered in monitoring P2P botnets in Sect. 2.4 where the challenges affect the accuracy of the crawl data (cf. Non-Functional Requirement 4).

2.3.3 Sensor Nodes

Due to the presence of network devices that allow sharing of IP addresses across many machines, e.g., NAT, botmasters also experience the same problem of regular P2P networks which often have two distinct classes of devices: superpeers and non-superpeers. As peers behind NAT are not directly reachable, botnets follow a two-tier network structure to enable all bots participating in the overlay management remain connected among themselves. The non-superpeers rely upon the superpeers to remain connected to the botnet and to receive new updates or commands from the botmaster. Commands from the botmaster is then retrieved by polling the superpeers for newer updates. The superpeers can relay any information from the botmasters to requesting bots, and therefore circumventing the NAT traversal issues.

Fortunately, this two-tiered network design can be exploited in monitoring botnets. Kang et al. were the first to propose a mechanism called *sensors* to enumerate *structured* P2P botnets [13], e.g., Storm. The sensors are directly routable and are deployed using strategic DHT IDs intended to intercept route requests of other bots. Since the requests were initiated by the bots themselves, the sensors can identify the non-superpeers based on the intercepted request messages. In contrast to crawling, sensors can enumerate both superpeers and non-superpeers (cf. Sect. 2.3.2). This idea has also been extended and applied for monitoring other unstructured P2P botnets [21].

By exploiting the node announcement mechanism that is required in each P2P botnet, a sensor node can be *announced* to existing superpeer bots (cf. Sect. 2.2) using the *announceMsg* method. When a non-superpeer requests additional neighbors from a superpeer, information about the sensor node may also eventually be handed out. Hence, non-superpeers will include the sensor into their NL and thereafter will regularly probe the sensor for its responsiveness. Therefore, sensors can enumerate bots that are not directly routable or discovered via crawling. An example of the depiction is provided in Fig. 2.3.

As explained in Sect. 1.1.3, entries in an NL are only removed or replaced if the associated bot has (consistently) remained unresponsive when being probed. To avoid being removed from the NL of the bots, sensors must always be responsive when being probed by bots. The high availability of a sensor also directly influences

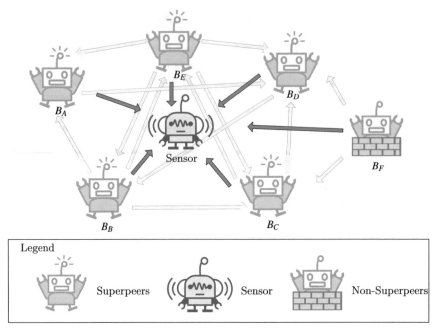

Fig. 2.3 A sensor enumerating both superpeers and non-superpeers

its popularity [15], i.e., the number of bots that have the sensor in their NL. The longer a sensor remains responsive, the higher the probability of bots propagating the sensor's information to other bots which in turn improves the coverage of the sensor.

A variation of a sensor node is often used in *sinkholing* attacks on P2P botnets. However, such attacks requires a vulnerability within the botnet protocol that is exploitable to overwrite information in a bot's NL. By invalidating all entries of a bot's NL, except those of the sensor(s), bots can only communicate to the sensor. As such, all communication from the botmaster can be filtered and thus the communication between the botmaster and its bots is disrupted.

Although sensors can enumerate bots that are not discoverable by crawlers, unfortunately, they cannot retrieve the connectivity information among bots. It is not possible to actively request the NL of the non-superpeers, except in cases where a botnet's design allows UDP hole-punching techniques [21]. Even though crawlers and sensors have their own set of advantages and disadvantages, they often used to complement each other to obtain better monitoring results. For instance, some researchers have augmented their sensors with crawling capabilities for monitoring [31, 32]. The following section looks at the challenges associated with monitoring activities.

2.4 Challenges in Botnet Monitoring

Botnet monitoring can be conducted easily by anyone with sufficient knowledge of the botnet's protocol. However, some challenges need to be considered while monitoring to ensure the reliability and the quality of the collected data. These challenges, if not addressed, can result in distorted or incomplete monitoring data that lead to wrong interpretations. (Non-Functional Requirement 3–5.)

These challenges are detailed in the following section and are mainly caused by the dynamic nature of P2P (bot-)networks itself (Sect. 2.4.1), noise resulting from monitoring activities by unknown third parties (Sect. 2.4.2), and anti-monitoring countermeasures that were deployed by the botnets (Sect. 2.4.3).

2.4.1 The Dynamic Nature of P2P Botnets

The dynamic nature of the P2P botnet overlay, which is similar to a regular P2P file-sharing networks, poses several challenges to monitoring mechanisms. These challenges are discussed in detail in the following:

1. **Churn and Diurnal Effects**: A botnet overlay experiences high *churn* rate of nodes joining and leaving the network at high frequency [26]. Therefore, crawlers that crawl bots with either a lower frequency or taking longer to complete a full crawl may introduce a significant network *bias*, i.e., in considering bots to be online that have already went offline, within the produced snapshot. In addition, newly arrived peers might also be missed by the crawler [26].
   Bots within the overlay also experience *diurnal effects* where significant portions of bots go offline and come online based on geographical timezones [26], e.g., computers that are turned on/off during or after working hours. This observation also suggests that any short-term measurement of a botnet, i.e., less than a week, would be heavily influenced by such diurnal effects.
2. **IP Address Aliasing**: Primarily contributed due to the shortage of IPv4 addresses and security concerns, IP address *aliasing* frequently occurs in P2P botnets. ISPs and organizations use devices such as NAT and proxies to share their network's limited number of available IP addresses. However, the network traffic generated by several machines behind such a device would seem to originate from only a single IP address. Thus, measurements that rely upon IP addresses alone may *underestimate* the total number of infected machines.
   Moreover, ISPs or organizations that run a DHCP service for a dynamic allocation of IP addresses for their users may also affect ongoing measurements. For instance, traffic originating from an infected machine will be observed from several IP addresses due to different addresses (re)allocated to existing bots by the DHCP servers, e.g., after reboots or the expiry of a lease period. IP address aliasing can also occur due to the presence of load-balancing infrastructures that may be used by the network administrators. These aliasing issues may lead to over-estimation of the number of infected machines.

Table 2.1 UIDs of existing
P2P botnets that are
retrievable from crawling [21]

Botnet	UID
Kelihos V1	16 bytes
Kelihos V2	16 bytes
Kelihos V3	16 bytes
Miner	None
Nugache	(Not shared)
Sality V3	4 bytes (non-persistent)
Sality V4	4 bytes (non-persistent)
Storm	16 bytes
Waledac	20 bytes
ZeroAccess V1	(not shared)
ZeroAccess V2	4 bytes (not shared)
GameOver Zeus	20 bytes

One way to overcome the IP address aliasing issue is to use persistent and unique
botnet-specific identifiers (UID), if applicable, to enumerate and associate the
infections accurately [21]. As depicted in Table 2.1, although many botnets use
some kind of UIDs, not all of them can be used to uniquely distinguish the bots.
For instance, the UID of Sality is not reboot-persistent and hence unreliable to be
used to distinguish unique bots. Moreover in some extreme cases like the Miner
botnet, there is no UID at all, hence rendering attempts to obtain a more accurate
estimation of the botnet population difficult or impossible. Therefore, unique bots
should be at least distinguished using the combination of the IP address and port
number in the absence of a reliable UID.

2.4.2   Noise from Unknown Third Party Monitoring Activities

To handle the dynamic nature of P2P networks, mediation steps can be designed to
obtain a more accurate measurements. However, the second challenging aspect of
botnet monitoring is the presence of unknown third party monitoring activities. As
explained in Sect. 2.3, botnet monitoring activities generate a considerable amount
of noise that is imprinted into the botnet's footprint. When third parties are unknown,
footprints of their monitoring activities will, unfortunately, be attributed as belonging
to the bots [14]. For instance, consider the scenario if you are interested in conducting
a churn measurement in a particular botnet. As discussed in Sect. 2.3.3, sensors that
are deployed, aim to be highly responsive for a prolonged period to ensure high
popularity. However, the presence of sensor nodes with longer session lengths may
skew the churn measurements as most bots have significantly shorter session lengths
[10]. Hence, any derived churn model may not be representative of the real churn in
the botnet.

Kanich et al. monitored the Storm botnet and took the active pollution into account by leveraging upon a flaw within the botnet's ID generator [14] (cf. Sect. 2.3.2). The authors classified the different type of sensors that were deployed based on their responses to sent request messages [8]. Besides that, they also described how they had to stop relying on the nature of the botnet's participants and carefully handle all received request and response messages. Their crawler often crashed within the first few minutes of crawling the network due to the presence of many malformed packets and the ongoing pollution attack within the botnet which uses bogon source IP addresses. Therefore, they had to put in much engineering effort, to enable their crawler to be fault-tolerant and continue crawling successfully. However, they also warned about future monitoring mechanisms which may become more stealthy and indistinguishable from bots. As a consequence, stealthy monitoring mechanisms will directly influence the botnet's footprint and further taint any gathered monitoring results.

2.4.3 *Anti-monitoring Mechanisms*

The third challenge in monitoring P2P botnets is the most interesting aspect of them all: anti-monitoring mechanisms deployed by botmasters. Botnets are an important asset to their botmasters. Consequently, their malicious activities also attract the attention of researchers and law enforcement agencies.

In the past, botnet monitoring activities have resulted in several botnets to be successfully taken down [20, 21]. For these reasons, botmasters are aware of the monitoring activities and have equipped recent botnets with anti-monitoring mechanisms. Moreover, many additional countermeasures have been proposed in the academia to impede crawling and deployment of sensor nodes. However, to the best of our knowledge, none of these proposals have yet been seen to be adopted by existing botnets.

Anti-monitoring mechanisms can be classified into the follow categories: (1) *Prevention*, (2) *Detection* and (3) *Response*. The first category of anti-monitoring mechanisms aim to impede or prevent monitoring activities. The second category focuses on detecting ongoing monitoring activities and the last category addresses on actions towards detected monitoring activities.

2.4.3.1 Prevention

Anti-monitoring mechanisms in this category, which are commonly implemented in many of recent botnets, aim to impede monitoring activities by design. Nevertheless, a majority of them are focused only on impeding performance of crawlers. In the following, mechanisms targeting crawlers will first be discussed, and followed by those targeting sensors.

A majority of anti-monitoring mechanisms against crawlers are focused on the neighborlist return mechanism of the botnets (cf. Sect. 1.1.3) as detailed in the following:

1. **Restricted neighborlist replies**: Many P2P botnets restrict the size of the returned *NL* when being requested, by handing out only a small fraction of their overall neighbors. In addition, botnets also utilize custom neighbor selection strategy to decide which neighbors are to be picked and returned when being requested. For instance, bots in Sality returns only one random (but active) neighbor [9] and ZeroAccess [29] return the 16 most-recently probed neighbors when requested. Meanwhile, botnets such as the GameOver Zeus [1, 4] implement a mechanism that returns only ten neighbors that are "close" to a botnet-specific ID specified within the *requestL* message and the ID of a bot's neighbors. This restriction mechanism utilizes the *Kademlia*-like XOR-distance metric [18] to calculate the notion of *closeness* between two bots. However, despite the presence of such restriction mechanisms, crawling is still possible. In Sality and ZeroAccess, a crawler needs to send *requestL* messages continuously to all discovered bots until the results converge, i.e., no newly discovered bots [17, 31, 32]. Similarly for the GameOver Zeus, a crawler needs to repeatedly query bots for their neighbor lists by spoofing different IDs chosen randomly as reported by Rossow et al. [21]. Nevertheless, the restriction mechanisms of these botnets implies that crawlers can only achieve a limited accuracy and are not able to provably retrieve or discover the *complete* neighborlist of a bot. Hence, the accuracy of the obtained monitoring data may be low or poor.

2. *Ratbot*: A theoretical and DHT-based structured P2P botnet called *RatBot* was proposed by Yan et al. that returns spoofed non-existing IP addresses when requested to hinder attempts to enumerate the botnet [30]. This mechanism makes the crawling process difficult and inefficient due to additional nodes that do not respond. Ratbot can also lead to an overestimation of the botnet size, which may be a preferred feature for botmasters, e.g., publicity among potential clients. Although crawlers may still work within RatBot, the introduction of excessive noise to the monitoring data may adversely affect botnet takedown attempts.

3. *Overbot*: Starnberger et al. [24] proposed a botnet called *Overbot*, which does not disclose the information of other bots or the botmaster if compromised by security researchers. The idea of Overbot is to let the infected machines to communicate to the botmaster using DHT keys that are generated by encrypting a *sequence number* with the public key of the botmaster. Bots utilize the DHT space within an existing P2P file sharing network like Overnet to publish *intentions* to communicate with the botmaster. By deploying several *sensor* nodes that listen for and decrypt search *requests*, a botmaster can identify bot-originated requests and communicate to the infected machines individually. However, since the sensors are assumed to have a copy of the botmasters private key, they pose themselves as a single point of failure if any sensor is compromised. Moreover, since bots continuously search for keys, such a pattern and noise can easily raise suspicions.

4. **Rambot**: Focusing on unstructured P2P botnets, Hund et al. proposed *Rambot* which uses a credit-point system to build bilateral trust amongst bots and use it as a *proof-of-work* scheme to protect against exploitation of its neighborlist exchange mechanism [12]. All nodes including crawlers are required to complete some computationally intensive tasks like crypto-puzzles before neighborlists are returned. However, with the advancement of computing resources that are available today, this proof-of-work mechanism can be easily circumvented.

5. **Manual neighborlist update**: Wang et al. in [28] proposed to allow botmasters to manually update the neighborlist of *all* bots from time to time. However, the design of the botnet requires frequent interactions from botmasters to instruct bots to *report* to a specific node with the information of their respective neighborlists and IDs. The authors also provided several suggestions to overcome the problem of this proposal posing itself as a single point of failure. Nevertheless, the design of the proposed botnet requires the botmaster to actively participate in the management of the botnet. This design not only increases the risk of the botmaster being exposed but also do not scale.

Very few work is available on anti-monitoring mechanisms against sensors. Andriesse et al. reported that is often difficult to identify sensors compared to crawlers that are in nature more aggressive in monitoring [2]. The authors also mentioned that due to the passive characteristics of sensor nodes and the difficulty in distinguishing them from bots, sensors often remain undetected. However, aggressive sensor popularization strategies such as *Popularity Boosting* [32] can be easily detected by mechanisms to detect crawlers (cf. Sect. 2.4.3.2).

It is worth mentioning that there are mechanisms deployed in existing botnets that are presumably aimed at preventing potential sinkholing attacks. These attacks often require the complete neighborlists of existing bots to be invalidated or filled up with only sinkhole servers. Since the sinkhole servers are a variation of sensors, those mechanisms aimed at preventing such attacks are listed in the following.

1. **IP-based filtering**: Most botnets, including Sality and ZeroAccess, have IP address filtering mechanisms that prevent multiple sensors sharing a single IP to infiltrate a botnet. Therefore, this mechanism prevents an organization or person with a single IP address from carrying out sinkholing attacks on the botnet. GameOver Zeus implements a more strict filtering mechanism that enforces that there is only a single entry allowed for a /20 subnet [1].

2. **Local reputation mechanism**: Bots in Sality use a local reputation mechanism that tracks the behavior of their neighbors based on the valid replies received when being probed (cf. Sect. 3.2.2). This mechanism slows down the rate of deployment of sensor nodes throughout the botnet as bots in Sality prefer existing and responsive neighbors over new bots.

3. **High-frequency swapping of neighborlist entries**: ZeroAccess uses a very small MM-cycle interval (cf. Sect. 3.3.2) to ensure that all neighbors are probed and cycled at a high rate. As a consequence, on the one hand, a sensor node loses its popularity unless it continuously announces itself to bots in ZeroAccess [32]. On the other, sensor that continuously announce itself would incur a high overhead in terms of the amount of communication messages.

2.4.3.2 Detection

Anti-monitoring mechanisms make it more difficult to monitor botnets. However, since this field is an arms race between the researchers and botmasters, it is only a matter of time before workarounds to circumvent or tolerate such mechanisms are available [16]. Hence, it is important that a detection mechanism is also in place to detect monitoring activities.

In the following, detection mechanisms for crawlers are presented:

1. **Rate-limitation of requesting neighborlists**: GameOver Zeus introduced a simple rate-limitation mechanism to detect crawling activities [1]. For that, a bot keeps track of the number of messages from each observed IP address within a sliding window of 60 s. If any IP address contacts a bot more than six times within an observation period, the IP is flagged as a crawler and remediation actions like blacklisting (cf. Sect. 2.4.3.3) are immediately taken. It is believed that the threshold value is set high enough, i.e., >6, to take into account possible false positives due to multiple bots sharing the same IPs, i.e., effects of NAT. Crawling is still possible although it requires more effort and delays between successive crawls for the same node. However, significant network noise and bias is introduced in the resulting botnet snapshot as more time is required for obtaining it. To circumvent this detection mechanism, a distributed crawling from a pool of unique IP addresses is required. The available IP addresses can be rotated among the crawlers to allow parallel crawling of bots without triggering the detection mechanism.

2. **Collaborative detection mechanism**: Andriesse et al. proposed a crawler detection mechanism that detects protocol anomalies resulting from improper protocol (re)implementations [2]. However, this can be easily evaded by strictly following the protocol. In addition, the authors also proposed a crawler detection approach that uses multiple colluding sensors to detect a crawler. This approach correlates the number of sensor nodes being contacted by a node and classifies one as crawler if the number of contacted sensors exceeds a certain threshold, e.g., maximum neighborlist size. The authors evaluated this mechanism with a deployment of up to 256 sensors in GameOver Zeus and Sality. Hence, this detection mechanism can also be easily implemented by botmasters on a large scale to detect crawlers.

Andriesse et al. reported that sensors are a more stealthy monitoring mechanism than crawlers due to their indistinguishableness from bots [2]. Although many sensors are observed to be highly popular, i.e., known by many bots [3], this behavior is indistinguishable from popular bots. Not much work has been done in the scope of detecting sensors that are deployed in botnets. In contrast, as one of the major contribution of this book, Sect. 5 will introduce three mechanisms to detect sensor nodes.

2.4.3.3   Response

After the detection of an ongoing monitoring activity, a response can be initiated. There are several possible actions to be implemented:

1. **Static blacklisting**: Many botnets are shipped with a list of IP addresses of organizations known to monitor botnets. Based on this list, bots can refuse to communicate to requests originating from those addresses.
2. **Automated blacklisting**: GameOver Zeus deploys an additional variation of blacklisting mechanism. It does not only rely upon the botmaster to update the list of blacklisted IPs. The botnet uses a rate-limiting mechanism (cf. Sect. 2.4.3.2) to identify crawling activities and subsequently blacklist them automatically [1]. Take note that the blacklisted entries are locally maintained and is not propagated to other bots.
3. **DDoS attack**: A more aggressive response is DDoS attacks on the IP or network of the monitoring node. This has been observed with the botmasters of GameOver Zeus who retaliated to the sinkholing attempts of their botnet [4]. Similar observations also been reported by researchers on the *Storm* botnet [25].

2.5   Summary

This chapter introduced requirements to botnet monitoring and presented a formal model for botnets that is used in this book, i.e., Chaps. 3–5. In addition, this chapter thoroughly analyzed the state of the art in botnet monitoring mechanisms (cf. Sect. 2.3); namely honeypots, crawlers, and sensors. From the analysis, it can be concluded that although honeypots can be quickly deployed, their monitoring coverage are limited compared to the other mechanisms. In particular, crawlers can enumerate and capture the inter-connectivity of many bots but with a drawback of missing out a fraction of bots, i.e., those behind NAT. Meanwhile, sensors are more effective in enumerating bots but often fail in capturing the inter-connectivity information. Hence, for best results, both crawlers and sensors should be deployed by complementing each other.

However, many of the previously discussed monitoring mechanisms were found to be not compliant with most of the non-functional requirements proposed in Sect. 2.1.2. Table 2.2 provides a summary of the existing monitoring mechanisms with respect to the compliance of the proposed requirements (cf. Sects. 2.1.1 and 2.1.2).

This chapter has also presented a thorough discussion in Sect. 2.4 on the three major challenges that need to be considered in botnet monitoring. These challenges, if left unaddressed, could impede the effectiveness of a monitoring mechanism. This would also not comply to the non-functional requirements proposed in Sect. 2.1.2.

Firstly, the dynamic nature of P2P botnets introduces significant amount of noise which may skew measurement results due to the presence of churn, diurnal effects, and IP address aliasing issues. State of the art reported that the usage of

Table 2.2 Summary of P2P botnet monitoring mechanisms regarding their compliance with the requirements specified in Sect. 2.1. Checkmark symbols ✓ indicate the fulfillment of these requirements, crossing symbols ✗ indicate their non-fulfillment, average symbols φ indicate a partial fulfillment, and dash symbols - indicate not applicable

P2P botnet monitoring mechanisms

	Requirements	Honeypots	Crawlers						Sensors		
			Glovenet [27]	StormCrawler [11]	Stormdrain [14]	Nugache Crawler [7]	Rossow et al. [21]	SPTracker [32]	Kang et al. [13]	Rossow et al. [21]	SPTracker [32]
Functional	Generic	✗	φ	φ	φ	φ	✓	✓	φ	✓	✓
	Protocol compliance	✓	✓	✓	✓	✓	✓	✓	✓	✓	✓
	Enumerator	φ	✓	✓	✓	✓	✓	✓	✓	✓	✓
	Neutrality	✓	✓	✓	✓	✓	✓	✓	✓	✓	✓
	Logging	✓	✓	✓	✓	✓	✓	✓	✓	✓	✓
Non-Functional	Scalability	✗	✓	✓	✓	✓	✓	✓	✓	✓	✓
	Stealthiness	✓	-	-	-	-	✓	✓	-	✓	✓
	Efficiency	✗	✓	✓	✓	✓	✓	✓	✓	✓	✓
	Accuracy	✗	φ	φ	✓	✓	φ	φ	φ	✗	✗
	Min. overhead/Noise	✓	✓	✗	✓	✓	φ	φ	✓	✓	✓

high-frequency crawling in combination with a long-term crawling can minimize bias introduced by churn and diurnal effects in P2P botnets. In addition, bias from IP address aliasing can be further reduced (or eliminated) using UIDs of botnets; if the UIDs are persistent and unique.

Secondly, the noise introduced by unknown third party monitoring activities (cf. Sect. 2.4.2) could also distort the resulting monitoring data. This particular challenge has seen very little attention from the research community in the context of P2P botnets. Analysis of the state of the art indicated that most of the existing botnet monitoring mechanisms were implemented only on a *best-effort* basis, and, as such, may have had their measurements tainted by noise originating from unknown monitoring activities, e.g., abnormally high uptime of sensor nodes.

Thirdly, anti-monitoring mechanisms which impede the performance of the monitoring mechanisms (cf. Sect. 2.4.3) also pose itself as a major hurdle for botnet monitoring. Fortunately, most of the existing anti-monitoring mechanisms observed deployed in the wild are still in their infancy in terms of their effectiveness. Although some of them can be circumvented or tolerated by the existing monitoring mechanisms, it may just be a matter of time before more advanced countermeasures are implemented by botmasters to raise the stakes. Along that line, this book will propose advanced countermeasures from the perspective of a botmaster in Sects. 4.2 and 5.1 to anticipate the retaliation of the botmasters against botnet monitoring.

Concluding, crawlers and sensors seemed to be the only viable solution to monitor P2P botnets in an effective and efficient manner. However, more effort has to be taken to ensure the stealthiness, efficiency, and accuracy of the monitoring mechanisms is improved to obtain high quality monitoring data. Therefore, future monitoring mechanisms need to also carefully consider and address the various challenges discussed in Sect. 2.4 to ensure an effective botnet monitoring that can be useful for further steps such as botnet takedown attempts or malware cleanup campaigns.

References

1. Andriesse, D., Rossow, C., Stone-Gross, B., Plohmann, D., Bos, H.: Highly resilient peer-to-peer botnets are here: an analysis of Gameover Zeus. In: International Conference on Malicious and Unwanted Software: "The Americas" (2013)
2. Andriesse, D., Rossow, C., Bos, H.: Reliable recon in adversarial peer-to-peer botnets. In: ACM SIGCOMM Internet Measurement Conference (IMC) (2015)
3. Böck, L., Karuppayah, S., Grube, T., Mühlhäuser, M., Fischer, M.: Hide and seek: detecting sensors in P2P botnets. In: IEEE Conference on Communications and Network Security, pp. 731–732 (2015)
4. Polska, C.E.R.T.: Zeus-P2P monitoring and analysis. Technical report, CERT Polska (2013)
5. Dagon, D., Gu, G., Lee, C.P., Lee, W.: A taxonomy of botnet structures. In: Computer Security Applications Conference (ACSAC), pp. 325–339. IEEE (2007)
6. Davis, C.R., Neville, S., Fernandez, J.M., Robert, J.M., McHugh, J.: Structured Peer-to-Peer Overlay Networks: Ideal Botnets Command and Control Infrastructures?. Springer, Berlin (2008)
7. Dittrich, D., Dietrich, S.: Discovery techniques for P2P botnets. Stevens Institute of Technology CS Technical Report 4 (2008)

8. Enright, B., Voelker, G., Savage, S., Kanich, C., Levchenko, K.: Storm: when researchers collide. USENIX; Log. **33**, 6–13 (2008)
9. Falliere, N.: Sality: story of a peer-to-peer viral network. Technical report, Symantec (2011)
10. Haas, S., Karuppayah, S., Manickam, S., Mühlhäuser, M., Fischer, M.: On the resilience of P2P-based botnet graphs. In: IEEE Conference on Communications and Network Security (CNS) (2016)
11. Holz, T., Steiner, M., Dahl, F., Biersack, E., Freiling, F.: Measurements and mitigation of peer-to-peer-based botnets: a case study on storm worm. In: LEET (2008)
12. Hund, R., Hamann, M., Holz, T.: Towards next-generation botnets. In: European Conference on Computer Network Defense. IEEE (2008)
13. Kang, B., Chan-Tin, E., Lee, C.: Towards complete node enumeration in a peer-to-peer botnet. In: Proceedings of International Symposium on Information, Computer, and Communications Security (ASIACCS) (2009)
14. Kanich, C., Levchenko, K., Enright, B.: The heisenbot uncertainty problem: challenges in separating bots from chaff. In: Proceedings of the 1st USENIX Workshop on Large-Scale Exploits and Emergent Threats (LEET) (2008)
15. Karuppayah, S., Fischer, M., Rossow, C., Mühlhäuser, M.: On advanced monitoring in resilient and unstructured P2P botnets. In: IEEE International Conference on Communications (ICC) (2014)
16. Karuppayah, S., Roos, S., Rossow, C., Mühlhäuser, M., Fischer, M.: ZeusMilker: circumventing the P2P Zeus neighbor list restriction mechanism. In: IEEE International Conference on Distributed Computing Systems (ICDCS) (2015)
17. Kleissner, P.: Sality. In: Botconf (2015)
18. Maymounkov, P., Mazieres, D.: Kademlia: a peer-to-peer information system based on the XOR metric. Peer-to-Peer Systems. Lecture Notes in Computer Science, vol. 2429, pp. 53–65. Springer, Berlin (2002)
19. McCarty, B.: Botnets: big and bigger. Secur. Priv. IEEE **1**(4), 87–90 (2003)
20. Nazario, J.: Botnet tracking: tools, techniques, and lessons learned. Black Hat (2007)
21. Rossow, C., Andriesse, D., Werner, T., Stone-gross, B., Plohmann, D., Dietrich, C.J., Bos, H., Secureworks, D.: P2PWNED: modeling and evaluating the resilience of peer-to-peer botnets. In: IEEE Symposium on Security and Privacy (2013)
22. Salah, H., Strufe, T.: Capturing connectivity graphs of a large-scale P2P overlay network. In: IEEE International Conference on Distributed Computing Systems Workshops (2013)
23. Spitzner, L.: The honeynet project: trapping the hackers. Secur. Priv. IEEE **1**(2), 15–23 (2003)
24. Starnberger, G., Kruegel, C., Kirda, E.: Overbot: a botnet protocol based on Kademlia. In: 4th International Conference on Security and Privacy in Communication Networks. ACM (2008)
25. Stewart, J.: Storm worm DDoS attack (2007)
26. Stutzbach, D., Rejaie, R., Sen, S.: Characterizing unstructured overlay topologies in modern P2P file-sharing systems. In: ACM SIGCOMM Internet Measurement Conference (IMC) (2005)
27. Wang, B., Li, Z., Tu, H., Hu, Z., Hu, J.: Actively measuring bots in peer-to-peer networks. In: International Conference on Networks Security, Wireless Communications and Trusted Computing, vol. 1 (2009)
28. Wang, P., Sparks, S., Zou, C.C.: An advanced hybrid peer-to-peer botnet. IEEE Trans. Dependable Secur. Comput. **7**(2), 113–127 (2010)
29. Wyke, J.: The zeroaccess botnet mining and fraud for massive financial gain. Sophos Technical Paper (2012)
30. Yan, G., Chen, S., Eidenbenz, S.: RatBot: anti-enumeration peer-to-peer botnets. Information Security. LNCS, vol. 7001. Springer, Berlin (2011)
31. Yan, J., Ying, L., Yang, Y., Su, P., Feng, D.: Long term tracking and characterization of P2P botnet. In: International Conference on Trust, Security and Privacy in Computing and Communications, pp. 244–251. IEEE (2014)
32. Yan, J., Ying, L., Yang, Y., Su, P., Li, Q., Kong, H., Feng, D.: Revisiting Node Injection of P2P Botnet. Lecture Notes in Computer Science, vol. 8792. Springer International Publishing, New York (2014)

Chapter 3
The Anatomy of P2P Botnets

Most P2P-based botnets implement (MM) mechanisms, with different initialized parameters, to ensure bots remain connected among one another in a distributed manner. However, existing botnets often implement highly customized communication protocols and designs, hence these botnets need to be reverse engineered to fully understand their MM mechanism and the utilized communication protocols. The reverse engineering information is crucial for botnet monitoring mechanisms to interact with bots in a botnet.

Although there are many available resources describing the anatomy of existing P2P botnets, they are only described on a very coarse-grained level. Therefore, own reverse engineering work is often required to complement the existing literature for a better understanding of the anatomy of a botnet.

For the purpose of this book, three botnets were picked as case studies: GameOver Zeus, Sality, and ZeroAccess. These selected botnets are not only some of the most prevalent P2P botnets but also deployed anti-monitoring strategies as discussed in Sect. 2.4.3 to impede botnet monitoring. The first three sections of this chapter (Sects. 3.1, 3.2 and 3.3) describes the MM mechanism of GameOver Zeus, Sality, and ZeroAccess using the formal model presented in Sect. 2.2. In addition, botnet-specific details that are useful for discussion in the later part of this book are also highlighted based on own reverse engineering results.The results not only managed to validate the findings of other work on these botnets, but also provided new insights that were important foundation for the works presented in this book. Finally, Sect. 3.4 summarizes this chapter.

The original version of this chapter was revised: Belated corrections have been incorporated. The erratum to this chapter is available at https://doi.org/10.1007/978-981-10-9050-9_7

© The Author(s), under exclusive license to Springer Nature Singapore Pte Ltd., part of Springer Nature 2018
S. Karuppayah, *Advanced Monitoring in P2P Botnets*, SpringerBriefs on Cyber Security Systems and Networks, https://doi.org/10.1007/978-981-10-9050-9_3

3.1 Dissecting GameOver Zeus

GameOver Zeus or also known as *P2P Zeus* is a variant of the infamous banking trojan Zeus first observed in the wild around September 2011 [1]. Detailed technical descriptions of this botnet are available as published technical reports and scientific articles in [1–4].

In the following, Sect. 3.1.1 describes the bootstrapping process of bots in GameOver Zeus. Section 3.1.2 discusses about the membership management mechanism of this botnet. Finally, Sect. 3.1.3 introduces the blacklisting mechanism used by this botnet.

3.1.1 Bootstrapping Process

Upon infecting a new machine, the malware of GameOver Zeus generates a 160-bits unique identifier (UID) based on the hash value of the concatenated strings of the operating system's *ComputerName* and the *VolumeID* of the first hard-drive in the infected machine. Since the same UID is always reproducible as long as the mentioned variables do not change, the UID is persistent through reboots. This UID is stored and heavily used throughout the communication among other bots in GameOver Zeus as described in Sect. 3.1.2 and can be represented as a 40-hexadecimal characters string.

The bots are supplied with a *bootstraplist* embedded in their binary which consists of 50 entries of other existing infected machines or bots. The information in this list consists of a tuple of *IP Address*, *Port* number and a *UID* for each entry. After successfully infecting a machine, a bot utilizes this list to bootstrap itself into the botnet overlay as described in Sect. 3.1.2.3. This list also effectively becomes the initial NL of the bot as depicted as an example in Table 3.1 with a maximum of 50 entries, i.e., $NL^{MAX} = 50$.

3.1.2 Membership Maintenance Mechanism

Bots in GameOver Zeus carry out their maintenance activities periodically every 30 min. Within each MMcycle, i.e., MM-cycle, a bot probes for the responsiveness of its neighbors for up to five times using the *probeMsg* (see Sect. 2.2). Take note that

Table 3.1 Example of a GameOver Zeus bootstrap/neighborlist

No	IP Address	Port	UID
1	123.100.12.201	25235	45d5f530d28f49...<truncated>
2	214.86.57.2	15687	c89d3abf771315...<truncated>
...	...	...	...
50	150.80.86.87	29001	d1649c62b94280...<truncated>

the *probeMsg* message is also commonly referred to as the *VersionRequest* message in other literature [3, 4] as the message is also used to query and exchange the latest botmaster update(s).

A valid response that is received to a sent *probeMsg* indicates that a particular neighbor is responsive, i.e., being online.If a neighbor remained unresponsive for five consecutive attempts, it is discarded from the NL and the probing process is continued with the next entry in the NL. If a bot has less than 50 responsive entries at the end of the MM-cycle, it looks into a queue that contains information about the senders of unsolicited request messages, i.e., request messages sent by other bots during their MM-cycle, that were successfully processed by the bot. The bot sends a probe message to each of the candidate that is not already in the NL.If a valid response is received and the NL is not full, i.e., $|NL_v| < NL^{MAX}$, the bot is added into the NL.

In addition, another mechanism to refresh the NL kicks in after every sixth MM-cycle or 180 min if the NL is low on entries, i.e., $|NL_v| < 25$. After considering all the senders of unsolicited requests, the bot actively requests for new neighbors from its responsive neighbors using the message *requestL* (see Sect. 2.2). Bots in GameOver Zeus include their UID or *key s* in every sent *requestL* message. Upon receiving such a request, a bot replies the message by returning ten entries from their NL, which are selected based on a *neighbor selection criteria* as described in the following.

3.1.2.1 Neighbor Selection Criteria

Bots that need information about other bots in the network use the *requestL(s)* method to request neighbors, where *s* is the key of the requesting bot. Take note that *s* can also be generated or spoofed, as long as it is a valid key, i.e., a 160-bit key. On receiving a valid NL request, a bot returns a subset of its NL of size *l*, in GameOver Zeus usually $l = 10$, which are closer to key *s* of the query with regards to the Kademlia-like XOR-distance using the method *processRequestL(s)* [5]. As detailed in Algorithm 3.1, the queried node replies as follows: it first constructs a list *L* containing up to the first ten elements listed in its NL *NL* (Line 2). Then, it iterates over all remaining elements in *L* (Line 3). The key of each element $L[i]$ is compared to the elements of *NL* under consideration (Line 5). As soon as the algorithm finds an element $L[i]$ with a *smaller* XOR-distance to *s* than $NL[j]$, $NL[j]$ is replaced with $L[i]$ (Line 6).

Therefore, entries or keys with closer XOR-distance are more likely to be returned (Line 5), but only the entry with the *closest* key to *s* is *guaranteed*. The algorithm may not return the second-closest key if it is the first element in the initial list *NL* (cf. Line 2); in this case, if the closest key is stored at an index larger than 9, it will be definitely compared to the second largest key at index 0 and, used as a replacement for the second largest key at index 0, and not considered further. Obviously, other constellations may exist which may lead to some of the closest-ten nodes to be discarded.In summary, the algorithm will return close-by nodes but not necessarily the ten closest nodes. The order of the entries stored in a bot's NL is non-deterministic,

---

Algorithm 3.1: *processRequestL(s)*

---

1 **for** $i = 0; i < l \;\&\&\; i < |NL|; i + +$ **do**
2     $L[i] \leftarrow NL[i]$
3 **end**
4 **for** $i = l; i < |NL|; i + +$ **do**
5   **for** $j = 0; j < l; j + +$ **do**
6   **if** $XOR(NL[i], s) < XOR(L[j], s)$ **then**
7           $L[j] \leftarrow NL[i]$ **break**
8   **end**
9   **end**
10 **end**
11 **return** L

---

e.g., they can be sorted by the XOR-distance of the neighbor to the bot or by the timestamp an entry was last updated.

This particular neighbor selection criterion introduces bias in the entries within a bot's NL towards a its key. Although this selection criterion is similar to the DHT implementation in Kademlia, GameOver Zeus remains an unstructured P2P botnet.

3.1.2.2 Inserting New Entries

Whenever a bot decides to add new neighbors, each new neighbor candidate goes through a two-step *sanitization* phase (for the newest variant of GameOver Zeus). The candidates need to satisfy the following conditions:

1. **Port Range**: A candidate's source port is required to be within the range $10,000 - 30,000$.
2. **Sub-network Range**: Only one IP entry is allowed in the NL for every $/20$ sub-network.

The algorithm discards candidates that failed to fulfil any of the conditions and appends the rest in the NL.

3.1.2.3 Node Announcement and Update of Existing Entries

Since a botnet needs to include new infections within the botnet overlay, the botnet MM uses the *announceMsg* to announce the arrival of the new bots to others in the overlay (see Sect. 2.2). In the case of GameOver Zeus, the feature of a *announceMsg* is incorporated within the *probeMsg* sent by bots. A bot that receives a valid *probeMsg* will consider the sender to be added directly into the NL if the sender is not already included and the NL is not full. Alternatively, if the NL is full, the bot adds the sender to a queue which consists of potential candidates when the size of the NL falls below a threshold value of $|NL| < 25$. Therefore, when a newly infected machine attempts to probe entries in the bootstraplist (see Sect. 3.1.1), the bot possibly also inserts itself as a potential candidate for consideration. If a bot inserts this newly infected machine in its NL, information about the new bot will be propagated further when other bots request for neighbors.

To address the IP address aliasing issue (see Sect. 2.4.1), GameOver Zeus uses a reboot-persistent UID to uniquely identify bots. This UID is transmitted in a field within all communication messages of bots in GameOver Zeus. Whenever a bot receives a *probeMsg* that consists of a UID already known to the bot (in the NL) but with a different IP address and/or port number, the bot updates the entry with the new information. This update mechanism ensures that a bot eventually gets to update the entries of neighbors that rejoin the network with new IP addresses or port numbers.

3.1.3 Blacklisting Mechanism

GameOver Zeus implements a two-fold blacklisting mechanism in the bots to deter crawling activities of known and unknown researchers. Firstly, all bots have a static list of IP addresses that they refuse to communicate to, e.g., security organizations known to perform botnet monitoring [3]. Secondly, bots also implement a simple local rate-limiting mechanism to detect and blacklist crawlers [3]. For this, a bot that receives more than six request messages from a single IP address within a sliding window of 60 s, automatically adds the IP address into a list of blacklisted IP addresses and stop communicating [1]. This list is maintained locally and not exchanged with other bots.

3.2 Dissecting Sality

Sality is a botnet family that propagates through file-infection that has been around since mid-2003 [6]. This family has evolved from traditionally communicating with the botmaster via emails, to a complete P2P-based communication in early 2008 [6]. Based on the initial reporting on the P2P variant of Sality, there could have been up to four versions of this P2P botnet.

The first version of P2P Sality observed in the wild transmitted "Version 2" in its communication messages. Around early 2009, Sality version 3 has been first seen and said to be the largest variant of the P2P Sality [6]. This variant still remains active at the time of writing. Falliere reported that differences between the protocols implemented in version 2 and 3 are minimal [6]. Around late 2010, version 4 of this botnet was first seen.This variant introduces new features, leading to improved security and robustness, by addressing some of the weaknesses found in the earlier versions of the botnet. Nevertheless, most of the communication protocols of the botnets as mentioned earlier remain the same, except for the transmitted version number.

Since the information that is needed to understand the various work within the scope of this book is common across these different botnets, i.e., communication protocols, all variants of P2P Sality are henceforth referred to as Sality in the remainder part of this book, unless mentioned otherwise. Detailed technical description of this botnet for interested readers is available as published technical reports and scientific publications in [6, 7].

3.2.1 Bootstrapping Process

Upon infecting a new machine, a Sality malware starts to listen on a (UDP) socket for incoming request messages following the node announcement procedure explained in Sect. 3.2.2.1. This socket or port number is derived based on the operating system's `ComputerName` using a simple built-in algorithm. Bots in Sality utilize UIDs for some of the inter-bot communication messages but the UIDs are not persistent over reboots. Bots in Sality obtain their UID through a process that involves an existing superpeer helping to assign it (explained later in Sect. 3.2.2). The assigned UIDs are integer values between $0 - 2.0 \times 10^7$. In the bootstrap phase, this UID is initialized to a default value of 0.

Next, the bootstraplist which is passed on from the previous file infector, i.e., bot, is used to initialize the NL of the new bot. This list typically consists of up to 1000 entries. The bot's NL can hold up to a maximum of 1000 entries, i.e., $NL^{MAX} = 1000$, and has a structure as depicted in Table 3.2. Initially, the values for all fields in this list is set to a default value of 0. Then, the bot copies the *IP Address* and *Port Number* of entries from the bootstraplist into the NL. After initialization of the NL, the bot executes the bootstrapping process by invoking the first MM cycle. The following section details the MM as well as the purpose of the different fields within the NL of a bot.

3.2.2 Membership Management Mechanism

Each bot in *Sality* utilizes a MM to ensure connectivity amongst bots with an interval of 40 min. Within each MM-cycle, a bot v probes all neighbors in its NL_v, sequentially by executing three different but inter-related processes.

Table 3.2 Example of a Sality bot's NL

No	IP Address	Port	UID	GoodCount	LastOnline
1	123.100.12.201	25235	1.8×10^7	65	0
2	214.86.57.2	15687	1.9×10^7	45	2356
...	...	...	...	...	...
1000	150.80.86.87	29001	1.1×10^7	21	5561

Firstly, the bot probes the responsiveness of a neighbor using the *probeMsg* method (cf. Sect. 2.2) which utilizes a Sality-specific *Hello* message. When a bot receives a valid response to the sent *Hello* message, the neighbor's *LastOnline* value is set to the *current* timestamp. Otherwise, the value is set to zero if an invalid reply is received or the request timed out, before the next entry in the NL is probed. The *LastOnline* is widely used within Sality as a flag to indicate that a particular neighbor was responsive within the last MM-cycle. For each successful verification of a neighbor's responsiveness, the *GoodCount* of the corresponding entry is incremented by one. Similarly, when a timeout occurred, or an invalid reply is received, the bot decrements the neighbor's *GoodCount* accordingly. Over time, neighbors that are often responsive have a higher *GoodCount* value compared to those that are not.

The *Hello* message exchange is also leveraged to ensure the latest command from the botmaster is disseminated to all bots. A command comes in the form of a digitally signed and encrypted file which is also known as a *URLPack* [6]. An *URLPack* consists of a list of URLs which usually host additional malicious binaries that needs to be downloaded and executed frequently by the bots in their local machine, i.e., approximately twice every hour. Within each *Hello* message, the newest *sequence number* of the *URLPack* known to a bot is always transmitted.

By comparing the sequence number transmitted in the received messages, bots can update themselves with the latest update from the botmaster. Consider the scenarios where Bot_x is sending a *Hello* message to Bot_y:

1. Bot_x **has an older *URLPack* than** Bot_y: Upon inspecting the received message, Bot_y will notice that Bot_x has an older *URLPack* installed. Therefore, Bot_y responds to the message by attaching the latest *URLPack* which would later be applied by Bot_x.
2. Bot_x **has a newer *URLPack* than** Bot_y: Upon inspecting the sequence number of the received message, Bot_y will notice that Bot_x has a newer *URLPack* installed. Therefore, Bot_y responds with a message by stating the sequence number of its currently installed *URLPack*. Upon receiving the message and noticing that Bot_y has an older *URLPack*, Bot_x sends an additional *Hello* message to Bot_y that attaches the latest *URLPack*. Bot_y can then apply the newer update accordingly.
3. **Both bots have the same *URLPack***: No steps are taken if both bots have the same *URLPack*.

After verifying the responsiveness of a neighbor, the bot checks if it has a status of either a superpeer or non-superpeer (see Sect. 3.2.2.1). Bots in Sality use the default UID value of $UID = 0$ as an indicator that a bot's status has not been tested. Similarly, a non-zero value indicates that a bot's capability is tested and this second process is omitted. Section 3.2.2.1 elaborates this testing process in detail.

If the number of entries in the NL is low, i.e., $|NL_v| < 980$, at the beginning of the MM-cycle, a Sality-specific *Neighborlist Request* message (NL_{Req}) is sent to the responsive neighbor using the method *requestL* (cf. Sect. 2.2). Bots receiving an NL_{Req} will respond with a *Neighborlist Reply* (NL_{Rep}) message containing information of one randomly picked bot from a list of *only* responsive neighbors. For this, a temporary list is first constructed from the main NL but consisting of only entries that have non-zero *LastSeen* values. Upon receiving an NL_{Rep} reply message, the returned entry can be considered as a potential candidate for the bot's NL as elaborated in Sect. 3.2.2.2. It is also worth noting that all entries within a Sality's NL are only superpeers. After completing the three processes for the picked neighbor, the next neighbor is probed until all neighbors within the NL have been probed.

After all neighbors are probed, a bot conducts an additional clean-up step on the NL if the size of the list was at least 500 at the beginning of that particular MM-cycle. This clean-up process discards all entries that have low *GoodCount* values, i.e., *GoodCount* < -30, or UIDs that are not within a superpeer's assigned range, i.e., $UID < 1.6 \times 10^7$ (see Sect. 3.2.2.1).

3.2.2.1   Testing Superpeer Capability and Node Announcement

Since there is no centralized infrastructure in Sality, the testing of superpeer capability is done with the help of other existing superpeers within the overlay. For that, Bot_x first sends a *Node Announcement Request* message via the *announceMsg* method (cf. Sect. 2.2) to Bot_y, i.e., a responsive neighbor. Within the sent message, Bot_x includes information of which UDP port it listens for incoming unsolicited requests (cf. Sect. 3.2.1). Upon receiving this request message, Bot_y sends a *Hello* message to the IP address of Bot_x using the port specified within the received request message.

The premise of this decision is; if Bot_x is publicly reachable from the Internet, it should be able to respond successfully to the received *Hello* message. Hence, a valid response to the probe message indicates Bot_x's superpeer-capability and the failure to respond, e.g., timeout occurred, signifies its incapability. Based on the verification, Bot_y responds to the initial *Node Announcement Request* with a reply that includes a value: a random value between $[0 - 1.6 \times 10^7)$ when Bot_x is not reachable on the listening socket, i.e., non-superpeer, or $[1.6 \times 10^7 - 2.0 \times 10^7]$ otherwise. If Bot_x is identified as a superpeer candidate, Bot_y additionally adds Bot_x into its NL as well (cf. Sect. 3.2.2.2). This way, to further propagate the information about itself being a potential superpeer to other bots that may need additional neighbors, Bot_x relies only upon Bot_y.

Finally, upon receiving the *Node Announcement Reply* message from Bot_y, Bot_x uses the returned value as its own UID and omits to repeat this process in subsequent MM-cycles until the next reboot. However, in case a reply was not received from Bot_y, i.e., a timeout occurred, Bot_x sets its UID to zero and repeats this testing procedure with the next responsive neighbor.

3.2.2.2   Inserting New Entries

There are two different scenarios where a bot attempts to add a new neighbor: (1) through NL exchange when having a low number of neighbors (cf. Sect. 3.2.2) and (2) when testing a bot for superpeer capabilities (cf. Sect. 3.2.2.1). Algorithm 3.2 describes the method *insertNeighbor*() for both scenarios with an input parameter

---

Algorithm 3.2: *insertNeighbor(entry, isTested)*

---

1 **if** *NL.isFull()* && *isTested* ≠ *True* **then**
2 | *return* // NL is full
3 **end**
4 **if** *entry.IP* ∈ *NL.getAllIPs()* **then**
5 | **if** *NL[entry.IP].Port* ≠ *entry.Port* **then**
6 | | ·**if** *NL[entry.IP].Status* ≠ *Online* ‖ *isTested* **then**
7 | | | *NL[entry.IP].Port* = *entry.Port* // Update Port
8 | |
9 | *return* // Nothing else to do
10 **if** *NL.isFull()* && *isTested* **then**
11 | // Make room for a new entry
12 | *NL.popEntryWithLowestGoodCount()*
13 **end**
    // Append entry at the end
14 *NL.append(entry)*

---

entry, i.e., IP address and port number. To distinguish the different scenarios, Sality uses a flag parameter *isTested* to indicate if the *entry* is being considered after being tested for its superpeer capability or not.

First, the bot checks if the current NL is already full (Line 1) and returns if the method was invoked within Scenario 1. Otherwise, the algorithm proceeds and checks if the IP address of *entry* is already present within the *NL* (Line 3). In case the IP address is present, it is checked whether the corresponding port in the reply matches with the existing entry in the *NL* (Line 5). If the ports do not match, the bot additionally checks (Line 5) if the entry in the *NL* was marked *offline*, i.e., *LastSeen* = 0, during the last MM-cycle or if the method was invoked within Scenario 2. If either one of the conditions is satisfied, the old port of the entry is replaced with that in the *entry* (Line 6) and the method returns. This entry updating feature allows an existing bot that reappeared on a different port to be updated by existing bots by retaining the old entry along with its corresponding *GoodCount* value as well.

However, if the address is unknown and the method was invoked from within Scenario 2 when having a full *NL*, the bot additionally removes one entry from its NL that has the lowest *GoodCount* value to make room for this new candidate (Line 10). Finally, *entry* is appended to the end of *NL* (Line 11).

3.3 Dissecting ZeroAccess

ZeroAccess is a malware dropper family that distributes additional malware that focuses on financial fraud through pay-per-click (PPC) advertising [8]. The botnet utilizes *plugins* as dropped modules to enable bots conducting the above-mentioned malicious activities. As of early 2016, researchers have reported the discovery of

Table 3.3 Distinct botnets distinguished by ports in ZeroAccess *Version* 2

	Ports for 32-bit	Ports for 64-bit
Bitcoin mining	16464	16465
Click-fraud	16471	16470

two versions of ZeroAccess [4, 8]: *Version* 1 in May 2011 and *Version* 2 in April 2012. In contrast to the former that uses TCP protocol for communications between bots, the latter version adopts UDP protocol instead. In the newer version, the set of commands utilized for inter-communication has also been reduced. This includes the removal of a command that can be exploited to launch sinkholing attack on the botnet.The removal of the command enhanced the efficiency and resiliency of the botnet [8]. Since it was difficult to find active bots to bootstrap in the *Version* 1 of this botnet, the remainder part of this book focuses only on the *Version* 2 [4]. Take note that this botnet was partially sinkholed by Symantec in 2013, but remnants of it remained active until now [9].

ZeroAccess *Version* 2 primarily performs two types of malicious activities: Bitcoin mining and Click-Fraud. For each activity, there exist two separate networks of bots distinguished by the OS architecture of the infected machines, i.e., 32-bit or 64-bit. As such, there are four distinct networks of ZeroAccess *Version* 2 as detailed in Table 3.3. Each of the networks distinguishes itself by the usage of distinct UDP hard-coded ports for communications.

Since the information that is needed to understand the various work within the scope of this book requires only the understanding of the communication protocols that is common across these different networks, all networks of ZeroAccess *Version* 2 are henceforth referred to simply as ZeroAccess in the remainder part of this book. Detailed technical description of this botnet is available as published technical reports and scientific publications in [4, 8–10].

3.3.1 Bootstrapping Process

Upon infecting a new machine, a ZeroAccess malware listens on a fixed UDP socket for incoming request messages as depicted in Table 3.3. Bots in ZeroAccess also generate a UID upon initialization that is not persistent over reboots. Next, the bots download all available *plugins* that enable add-on features as intended by the botmasters, e.g., Click-fraud and Bitcoin mining, and store them on the infected machine [8].

According to the reverse-engineering results of a malware variant[1], the bots have three types of NLs: *primary*, *secondary* and *backup* list. The bot's *primary* NL can hold up to a maximum of 256 entries, i.e., $NL^{MAX} = 256$, and has a structure

---

[1]md5 = ea039a854d20d7734c5add48f1a51c34

Table 3.4 Example of a ZeroAccess *Primary* NL

Index	IP Address	LastOnline
0	123.100.12.201	2564
1	214.86.57.2	1082
...	...	..
255	150.80.86.87	220

as depicted in an example in Table 3.4. The *primary* NL only maintains the IP address of a neighbor as all bots listen on a dedicated port that is unique to the network they reside in. The bootstraplist of the malware that typically has up to 256 entries is used to initialize the *primary* NL of the new bot. The remaining lists which have a significantly larger length, i.e., 16×10^6 entries, are initialized empty during infection and information of all responsive and known bots throughout the lifetime of the particular bot is continuously added/updated in both lists. However, only the *backup* list is persistent over reboots. The presence of this list allows a bot to recover from any potential sinkhole attempts using bots that were responsive in the past (cf. Sect. 3.3.2). After initialization of the NLs, the bot executes the bootstrapping process by invoking the first MM cycle.

3.3.2   Membership Management Mechanism

Each bot in *ZeroAccess* utilizes a MM mechanism with an interval of 256 s. Within each MM-cycle, a bot v sequentially probes each neighbor in its *primary NL_v* and optionally two additional bots from the *secondary* and *backup* lists, every second using the *probeMsg* method (cf. Sect. 2.2). A bot starts probing entries from index $0 - 255$ each every second. Upon reaching the final entry, i.e., $index = 255$, the iteration process wraps up to start again from $index = 0$.

Maintenance of the NL in the bots relies on two types of message exchanged among bots: *getL* and *retL*. As explained above in Sect. 3.3.1, each bot listens on the botnet-specific port for unsolicited requests, i.e., *server port*, and sends such requests from an OS-allocated but fixed UDP port or a *client port*. As such, all probing messages originate *only* from the client port. Figure 3.1 depicts such a probing process between two bots: Bot_X and Bot_Y.

Let's assume Bot_X probes Bot_Y for its responsiveness. Upon receiving a *getL* message, Bot_Y responds with a *retL* message that consists of a subset of its own *primary* NL and a list of all *plugins* available for sharing, i.e., using TCP connections.

Fig. 3.1 Message exchange for *Bot$_X$* probing *Bot$_Y$* in *ZeroAccess*. *Source*: https://ieeexplore.ieee.org/stamp/stamp.jsp?tp=&arnumber=7510885

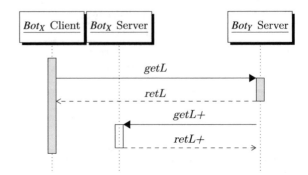

Section 3.3.2.2 discusses the details on the neighbor selection criteria utilized to pick the neighbors to be included in a reply. In addition to the reply, a probe message is sent to the *server port* of *Bot$_X$*, i.e., lower part in Fig. 3.1. This process which serves as a method to verify if *Bot$_X$* is a superpeer, requires *Bot$_X$* responding to the probe message of *Bot$_Y$* using the server port as the source port in the reply. A successful response indicates that *Bot$_X$* is routable, i.e., superpeer.

Algorithm 3.3: *processGetL(sender, msg)*

```
1 // Reply to all received requests
2 rep ← createRetL(msg.getFlag())
3 send(sender, rep) // Send a retL to sender
4 if msg.getFlag() == 0 then
5 │   // Send a getL+
6 │   send(sender, createGetL(flag = 1))
7 end
```

To avoid the exchange of messages to be continuously looped, a *flag* in the messages is set when being probed to indicate *no further probing is needed* as depicted in Algorithm 3.3. Such messages are differentiated by using *getL+* and *retL+* when referring to the set of messages that do not require additional probing.

Upon receiving either a *retL* or *retL+*, bots process the message by performing the following steps:

1. Inserts the sender in the *primary* NL as described in Sect. 3.3.2.1.
2. Inserts the sender and the returned neighbors (if any) in the *secondary* NL.
3. If there is, at least, one *plugin* information in the reply, inserts the sender in the *backup* NL.

Although the sender is added into all three NLs, for the remainder part of this book, only the *primary* NL is focused, which is used to select neighbors to be returned in *retL* and *retL+* messages. Take note that due to the communication design adopted by this botnet, researchers have reported that they were able to leverage them to conduct UDP hole-punching to communicate continuously to non-superpeers that first reached out to a sensor node [4].

3.3.2.1 Inserting New Entries

Upon receiving a valid response, a sender's IP is added or updated within the *primary* NL of each bot.For that, the IP is used as a parameter to invoke Algorithm 3.4 that is responsible for handling this process. Firstly, bots check if the sender's IP exists in their *primary* NL (Line 2). If the IP exists, this entry is removed from the NL (Line 4).After removing this entry, all subsequent entries are shifted up one position to bridge the gap originating from the deletion process. Finally, the sender's IP is used to create an entry at the beginning of the NL, i.e., $index = 0$. Afterward, the *LastSeen* value is set to the value of the current timestamp.

Algorithm 3.4: $insertInPrimaryList(sender)$

---

1 // Check if sender known
2 **if** $sender \in NL.getAllIPs()$ **then**
3 | // Remove existing entry and close the gap
4 | $NL.pop(sender)$
5 **end**
6 // Push entry at the beginning of list
7 $NL.push(sender)$
8 // Set sender's LastSeen to now
9 $NL[sender].LastSeen = getCurrentTimestamp()$

Due to the nature of the process of inserting neighbors, the *primary* NL of ZeroAccess bots is sorted by *most-recently* responsive neighbors. If the NL is full and a new (but non-existing) entry needs to be added, the last entry, i.e., $index = 255$, is discarded when the new entry is pushed at the beginning of the NL.

3.3.2.2 Neighbor Selection Criteria

Bots need to respond to each of received probe messages, i.e., *getL* and *getL+* messages, with a *retL* message that includes a subset of neighbors that have their responsiveness most recently verified. According to the protocol, a bot can include up to 16 entries in a resulting reply but an empty reply is also valid.Since the *primary* NL of the bots is always sorted by most responsive neighbors first, a bot simply returns the first-16 entries from the NL in the reply message.

3.4 Summary

This chapter described the anatomy of three P2P botnets: GameOver Zeus, Sality, and ZeroAccess, based on own work of malware reverse engineering. Particularly, the bootstrapping and MM mechanism of each botnet was thoroughly analyzed, and the differences among them were outlined. From the analysis, existing P2P botnets are found to share similarities on common MM designs and anti-monitoring mechanisms aimed at impeding botnet monitoring.

However, some of the parameters used by the botnets differed greatly among one another, e.g., MM-interval or the size of the NL used by the botnets, as described in the following:

- **MM-interval**: The MM-interval directly influences the rate of stale information in the NL of bots as well as the incurred communication overhead. For instance, a long interval may cause a bot to be isolated from the overlay due to many of its neighbors being non-responsive, e.g., bots gone offline. However, the communication overhead caused by such large intervals are lesser and is helpful to remain stealthy. Sality and GameOver Zeus utilized such intervals with 40 and 30 min respectively.
  In contrast, a short interval may ensure responsive neighbors are always retained in the NL. Therefore, chances of a bot with short interval to be isolated from the botnet overlay is much lesser than those with larger interval. However, a short interval also implies that higher communication overhead is incurred by frequently probing the responsiveness of the neighbors. This in turn may easily raise suspicions to network administrators. ZeroAccess is an example of such botnet with a short interval, i.e., 256 s.
- **Size of NL**: Ensuring a bot remains connected to the botnet overlay is also often influenced by the availability of sufficient number of responsive neighbors to communicate with. Moreover, the size of the NL utilized by botnets also influences how quickly a command issued by a botmaster is disseminated throughout the botnet. In the context of the analyzed botnets, GameOver Zeus utilized the smallest size for its NL, i.e., $|NL| = 50$. This is followed by ZeroAccess with a size of $|NL| = 256$ and Sality with the biggest NL size, i.e., $|NL| = 1000$.
- **Anti-monitoring mechanisms**: Each of the anti-monitoring mechanisms implemented by the analyzed botnets are unique and has its special purpose in protecting the botnet either from being monitored or taken down. GameOver Zeus is by far the most advanced botnet among the three botnets analyzed in this chapter. It utilized an NL restriction mechanism that returns only a subset of its NL following a special node selection criteria as described in Sect. 3.1.2.1. In addition, the botnet also uses IP address-based filtering to ensure sensors or sinkhole server entries are not able to quickly or easily fill up the NL of a bot. Finally, GameOver Zeus also utilizes blacklisting mechanisms to refuse communicating to known or aggressive crawlers.
  Sality also introduced an NL restriction mechanism that returns only one random entry out of the 1000 entries for each received NL request. Moreover, it also

implemented a local reputation mechanism whereby previously-known neighbors are preferred to new neighbors. This reputation mechanism prevents sinkholing attacks that aim to invalidate all entries easily within the NL of a bot.

Finally, ZeroAccess utilizes the short MM-interval to make it difficult for sinkholing attacks by cycling entries within the NL with a high frequency. Since most sinkholing attack requires the existing entries in the NL of a bot to be invalidated, this mechanism quickly flushes away the invalidated entries. As a consequence, a successful sinkholing attack requires a lot of resources to continuously invalidate the entries in the NL of all bots in the botnet. In addition to this design, the botnet also maintains two additional NLs on top of the main NL used for regular MM activities. These two NLs keep track of all responsive bots ever discovered since the last reboot of an infected machine. Entries within these NLs are also contacted from time to time to allow the botnet to recover even from a powerful ongoing sinkholing attack.

Many of the analysis presented in this chapter serve as a foundation to other works presented in this book. For instance, the neighbor selection criteria of GameOver Zeus that is presented in Sect. 3.3.2.2, is important to understand the work on circumventing this particular mechanism (see Sect. 4.1.1). In addition, the MM design of Sality and ZeroAccess as presented in this chapter is leveraged in the work on autonomously detecting crawlers in P2P botnets (see Sect. 4.2.2). Moreover, the communication and MM designs utilized by both Sality and ZeroAccess are also leveraged to present several novel sensor detection mechanisms in Chap. 5.

References

1. Andriesse, D., Rossow, C., Stone-Gross, B., Plohmann, D., Bos, H.: Highly resilient Peer-to-Peer botnets are here: an analysis of Gameover Zeus. In: International Conference on Malicious and Unwanted Software: The Americas (2013)
2. Abuse.ch: Zeus Gets More Sophisticated Using P2P Techniques (2011)
3. Polska, C.E.R.T.: Zeus-P2P monitoring and analysis. Technical report, CERT Polska (2013)
4. Rossow, C., Andriesse, D., Werner, T., Stone-gross, B., Plohmann, D., Dietrich, C.J., Bos, H., Secureworks, D.: P2PWNED: modeling and evaluating the resilience of Peer-to-Peer botnets. In: IEEE Symposium on Security and Privacy (2013)
5. Karuppayah, S., Roos, S., Rossow, C., Mühlhäuser, M., Fischer, M.: ZeusMilker: circumventing the P2P Zeus neighbor list restriction mechanism. In: IEEE International Conference on Distributed Computing Systems (ICDCS) (2015)
6. Falliere, N.: Sality: Story of a Peer-to-Peer Viral Network. Technical report, Symantec (2011)
7. Kleissner, P.: Sality. In: Botconf (2015)
8. Neville, A., Gibb, R.: ZeroAccess Indepth. Symantec Security Response (2013)
9. Symantec: Grappling with the ZeroAccess Botnet (2013)
10. Wyke, J.: The ZeroAccess BotnetMining and Fraud for Massive Financial Gain.Sophos Technical Paper (2012)

Chapter 4
Crawling Botnets

Crawlers are widely used in botnet monitoring (see Sect. 2.3.2) to enumerate bots and to discover the interconnectivity among them. Such information is vital to law enforcement agencies in botnet takedown attempts. In response, botmasters have introduced several anti-crawling mechanisms to impede monitoring activities. In particular, GameOver Zeus can be considered as the most sophisticated P2P botnet seen to date [1] due to the significant efforts that are taken to impede monitoring activities. As discussed in Sect. 2.4.3, such anti-crawling mechanisms can introduce a significant amount of noise and distortion to the data gathered by monitoring. Hence, it is important to circumvent these mechanisms to obtain monitoring data of better quality and to anticipate future botnet advancements.

Section 4.1 presents work on circumventing the NL restriction mechanism and the automated blacklisting mechanism of GameOver Zeus from an attacker perspective, i.e., researchers and legal enforcement agencies. Section 4.2 introduces advanced anti-crawling mechanisms that aim at impeding crawling activities and on the detection of ongoing crawling activities from the perspective of botmasters. Section 4.3 presents the evaluation results of the proposed mechanisms on the newly introduced countermeasures and the advanced crawling mechanisms. Finally, Sect. 4.4 provides a brief discussion and summarizes this chapter.

4.1 Circumventing Anti-crawling Mechanisms

In this section, the anti-crawling mechanism of GameOver Zeus (see Sect. 3.3.2.2) are thoroughly analyzed, and methods to circumvent them are presented. In addition, a novel crawling strategy is also proposed to improve the efficiency of crawlers.

The original version of this chapter was revised: Belated corrections have been incorporated. The erratum to this chapter is available at https://doi.org/10.1007/978-981-10-9050-9_7

S. Karuppayah, *Advanced Monitoring in P2P Botnets*, SpringerBriefs on Cyber Security Systems and Networks, https://doi.org/10.1007/978-981-10-9050-9_4

43

4.1.1 Restricted NL Reply Mechanism of GameOver Zeus

GameOver Zeus is dubbed a sophisticated botnet due to the effective mechanisms adopted to impede monitoring activities and to recover in a case of a potential take-down [1, 2]. One particularly interesting defense mechanisms is the NL restriction mechanism that deterministically picks and returns a subset of neighbors, specific to the requester, when being requested (see Sect. 3.1.2.1). This subsection will intro-duce a novel algorithm called ZEUSMILKER that circumvents this restriction mech-anism. For that, background information on the restriction mechanism is detailed in Sect. 4.1.1.1. Then, Sect. 4.1.1.2 introduces ZEUSMILKER algorithm.

4.1.1.1 Background

Bots in Gameover Zeus regularly exchange subsets of their NLs on a request basis to maintain and improve the connectivity of the botnet. The exchanged subsets are selected based on an *XOR-distance* metric between the sender's key included in request messages and the entries in the NL (see Sect. 3.1.2). Hence, two bots that request an NL from a bot may receive two different sets of entries. Thus, a botnet crawler has to query each node multiple times using distinct spoofed keys, which decreases the performance of a crawler considerably [3].

Rossow et al. first proposed a method to circumvent this mechanism by spoofing the querying keys randomly, hoping to obtain all neighbors [1]. In contrast, in the following, a reliable method to provably obtain all neighbors from a bot by strate-gically spoofing requester UIDs or keys is presented. This method is the only work known to provide solution in successfully circumventing the restriction mechanism of GameOver Zeus.

4.1.1.2 ZEUSMILKER Algorithm

The ZEUSMILKER algorithm that is presented in Algorithm 4.5 leverages the design of the NL restriction mechanism itself (see Sect. 3.1.2.1). First, important notations to understand the algorithm is introduced.

Notations

Each bot is assigned a unique *key* in the form of a b-bit string. $\mathbf{1}(i)$ is used to denote a bit string of i 1s and analogously $\mathbf{0}(i)$ denotes a string of i 0s. Furthermore, $|s|$ denotes the length of a string s, and $||$ is the concatenation operator. For two b-bit keys x and y, the function $cp(x, y)$ returns their common prefix. An order on the set of b-bit keys is defined by associating the key's bits $b_{b-1} \ldots b_0$ with an integer value $\sum_{i=0}^{b-1} 2^{b_i}$. In particular, a key y is defined bigger, smaller or equal than a key x by comparing the integer values. The operators $+$ and $-$ are then defined as the respective operators in $\mathbb{Z}$, the set of all integers. Two keys x and y are called *consecutive* if $y = x + 1 \mod 2^b$. Finally, $I(x, y) = \{x + 1, \ldots, y - 1\}$ is used to denote the set of all possible keys '*between*' x and y. Note that the set is empty if $y \leq x$.

ZEUSMILKER aims on retrieving the complete NL NL of a bot using the method *requestL(s)* as described in the formal model (see Sect. 2.2). Algorithm 4.5 achieves this goal by discovering pairs of keys (x, y) such that the NL is guaranteed not to contain any keys in $I(x, y)$ and thus $NL \cap I(x, y) = \emptyset$. The algorithm terminates if no sets of $I(x, y)$ can contain additional and yet unknown keys, guaranteeing that the list of returned keys L is identical with NL. The NL of bots is assumed to remain static during crawling. As GameOver Zeus has a rate-limiting mechanism (see Sect. 3.1.3) in place, several unique IP addresses are assumed to be utilized to circumvent this mechanism, i.e., crawling bots using multiple IP addresses.

The neighbor selection mechanism (see Sect. 3.1.2.1) returns ten entries that are close to the supplied key s through the method *requestL(s)*. The selection of the returned neighbors is based on a well-known XOR-distance metric that was introduced in Kademlia [4]. However, the design of the GameOver Zeus mechanism is such that only the closest key is always guaranteed to be returned. This observation along with the nature of an XOR operation is leveraged by ZEUSMILKER to circumvent the GameOver Zeus restriction mechanism. Before discussing Algorithm 4.5 in detail, a short explanation is provided on how spoofing with two consecutive keys $s_1, s_2 \in I(x, y)$ results in a set $I(x, y)$, such that all keys in $I(x, y)$ are *not* contained in NL.

Consider the left-hand side of Fig. 4.1: Here, all possible b-bit keys are represented in the form of a ring. Note that all the keys in the right half of the ring are closer to $\mathbf{0}(b)$ than $\mathbf{1}(b)$ with regard to the XOR-distance, whereas all keys on the left half are closer to $\mathbf{1}(b)$. Similarly, when considering only the keys on the right half, the keys in the upper right quarter are closer to $00||\mathbf{1}(b-2)$ than to $01||\mathbf{0}(b-2)$, whereas the keys in the lower right quarter are closer to $01||\mathbf{0}(b-2)$.

In this manner, one can successively divide keys into sets according to their closeness in their XOR-distances to the supplied or spoofed keys. This division is leveraged to identify keys *not* contained in the NL NL and the keys *possibly* contained in NL as follows. Let

$$s_1 = c||0||\mathbf{1}(i), \quad s_2 = s_1 + 1 = c||1||\mathbf{0}(i) \tag{4.1}$$

for some common prefix c and $i \geq 0$, i.e., s_1 is a key ending with a string of 1s, and s_2 is the next higher key, thus ending with a string of 0s. First note that for any keys id_1 and id_2, $XOR(id_1, id_2)$ starts with a string of 0s of the length of their common prefix. So, if id_1 shares a longer common prefix with id_2 than with a key id_3, id_1 is closer to id_2 than to id_3 with regard to the XOR distance. Now, assume that the NL contains keys k_1 and k_2 starting with $c||0$ and $c||1$, respectively. As a consequence,

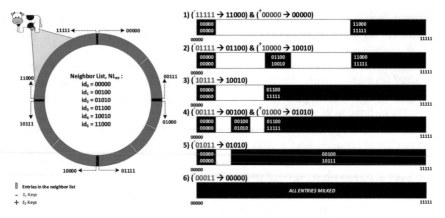

Fig. 4.1 Visual representation of the key space (Example 27). The '+' keys discover the next bigger key, whereas the '−' keys reveal the next smaller key. *Source*: https://ieeexplore.ieee.org/stamp/stamp.jsp?tp=&arnumber=7164947

x and y, the closest keys in NL to s_1 and s_2 with respect to the XOR distance, have to start with $c||0$ or $c||1$, respectively. So, $requestL(s_2)$ returns a list containing a key $y = c||1||r_y = s_2 + r_y$ for some i-bit string r_y. Similarly, $requestL(s_1)$ returns a list containing a key $x = c||0||r_x = s_1 - \mathbf{1}(i) + r_x$ for some r_x. Therefore, it is shown that indeed x and y are such that $NL \cap I(x, y) = \emptyset$. By the definition of $I(x, y)$, $I(x, y) = I(x, s_2) \cup I(s_1, y)$. The claim that $NL \cap I(x, y) = \emptyset$ follows from showing that all $z_x \in I(x, s_2)$ and $z_y \in I(s_1, y)$ have a lower XOR distance to s_1 or s_2 than x or y, respectively, and hence cannot be contained in NL. Note that all $z \in I(x, y)$ share the prefix c. Consider $z_y = c||1||q \in I(s_1, y)$ for an i-bit string q, so that $XOR(z, s_2) = q = z - s_2$. As a consequence, $XOR(z_y, s_2) < XOR(y, s_2)$ for all keys $z_y \in I(s_1, y)$, so that $z_y \notin NL$ if y is the closest key to s_2 in NL. Similarly, for any $z_x = c||0||q \in I(x, s_2)$, $r_x < q \le \mathbf{1}(i)$, so that $XOR(z_x, s_1) = \mathbf{1}(i) - q$ and hence $XOR(z_x, s_1) < XOR(x, s_1)$. Hence, $z_x \notin NL$ if x is the closest returned key to s_1. In summary, all keys in $I(x, y)$ are not contained in NL, and thus a method to reliably identify sets of keys that are guaranteed not to be contained in NL is found. However, without further queries, it is not possible to say which keys in $I(k_1, x)$ and $I(y, k_2)$ are contained in NL.

Example 4.1 As an example consider the neighborlist
$NL_{ex} = \{00000, 00100, 01010, 01100, 10010, 11000\}$ and assume for simplicity that each query via $requestL()$ only returns $l = 1$ key. Assume it is already discovered that $k_1 = 00000$ and $k_2 = 01100$ with common prefix $c = 0$. The next step is to query with $s_1 = 0||0||111 = 00111$ and $s_2 = 01000$. $requestL(s_1)$ is guaranteed to return $x = 00100$ and $requestL(s_2)$ returns $y = 01010$. However, the reply does not tell if any keys in $I(k_1, x) = \{00001, 00010, 00011\}$ or $I(y, k_2) = \{01011\}$ are contained in NL_{ex}.

Algorithm 4.5: ZeusMilker()

```
   // Initialization
1  L ← ∅                                              // Crawled keys
   // Get smallest key
2  M ← requestL(0(b))
3  L ← L ∪ M
4  k_first ← getClosestKey(M, 0(b))
   // Get largest key
5  M ← requestL(1(b))
6  L ← L ∪ M
7  k_last ← getClosestKey(M, 1(b))
8  if k_first ≠ k_last && k_first ≠ k_last − 1 then
9  │   R.push((k_first, k_last))                      // Push undiscovered range

   // While not fully discovered
10 while not R = ∅ do
       // Get keys for spoofing
11  │   (k_1, k_2) ← R.pop()
12  │   c ← getCommonPrefix(k_1, k_2)
13  │   s_1 ← c||0||1(b − length(c) − 1)
14  │   s_2 ← c||1||0(b − length(c) − 1)
       // Execute queries and add new sets
15  │   if k_1 < s_1 then
16  │   │   M ← requestL(s_1)                          // query with s_1
17  │   │   L ← L ∪ M
18  │   │   x ← getClosestKey(M, s_1)
19  │   │   if x ≠ k_1 then
20  │   │   │   R.push((k_1, x))

21  │   if k_2 > s_2 then
22  │   │   M ← requestL(s_2)                          // query with s_2
23  │   │   L ← L ∪ M
24  │   │   y ← getClosestKey(M, s_2)
25  │   │   if y ≠ k_2 then
26  │   │   │   R.push((y, k_2))

27 return L
```

Algorithm 4.5 now subsequently identifies sets of keys which cannot be contained in NL, while at the same time finding new keys k_1 and k_2 that are used for determining the keys s_1 and s_2. Initially, the list of discovered keys L is empty (Line 1). Then $s_1 = 0(b)$ and $s_2 = 1(b)$ are used as keys for the first two queries with the returned list $requestL(s_1)$ and $requestL(s_2)$ added to the set of discovered keys (Lines 2–7). In particular, $requestL(s_1)$ has to contain the smallest key k_{first} and largest k_{last} in NL, i.e., the closest keys to $0(b)$ and $1(b)$. Hence, the set $I(k_{last}, k_{first})$ is the first detected set of keys that are not contained in NL. However, $I(k_{first}, k_{last})$ potentially contains undiscovered keys, given that it is non-empty, i.e., the two keys are not equal or consecutive. So, the pair (k_{first}, k_{last}) is the first element in R (Line 9), which is implemented as a queue. Hence, R contains pairs (k_1, k_2) whose common prefix defines the spoofed keys in future iterations. In each iteration of the while loop

(Lines 10–26), such a pair (k_1, k_2) from the front of the queue is considered. The common prefix c of k_1 and k_2, determines the two spoofed keys s_1 and s_2, such that $s_1 = c||0||1(b - \text{length}(c) - 1)$, which consists of the common prefix c, 0, and a string of 1s achieving a total length of b, is the largest key closer to k_1 than to k_2 (in terms of the XOR-distance). Analogously, $s_2 = c||1||0(b - \text{length}(c) - 1) = s_1 + 1$ is the smallest key closer to k_2 than k_1 (Lines 12–14). If s_1 is not bigger than k_1, $I(k_1, s_1)$ is empty, hence it is not necessary to query with s_1. Analogously, if s_2 is not smaller than k_2, $I(s_2, k_2)$ is empty. If s_1 is bigger than k_1, the method call $requestL(s_1)$ is executed, the returned list M added to L, and the key x is chosen as the closest key to s_1 in M (Lines 16–18). Similar, if s_2 is smaller than k_2, y is chosen as the closest key to s_2 in the set returned by $requestL(s_2)$ (Lines 22–24). As discussed above, keys in $I(x, y)$ are guaranteed to not contained in NL, hence only the sets $I(k_1, x)$ and $I(y, k_2)$ can contain undiscovered keys if they are non-empty. Hence, the pairs (k_1, x) and (y, k_2) are added to the end of R (Line 20 and 26, respectively).

Example 4.2 The exemplary neighborlist $NL_{ex} = \{00000, 00100, 01010, 01100, 10010, 11000\}$ from Example 4.1 is used, which is sorted for simplicity and indexed by $id_j = NL_{ex}[j]$, for $j = 0 \ldots 5$. The ring on the left of Fig. 4.1 depicts how these keys map onto the whole key space. For simplicity, it is assumed that only $l = 1$ keys are returned per query. However, for larger $l' < |NL_{ex}|$, the same number of steps is required to *guarantee* that all keys in NL_{ex} are returned, though individual keys might be discovered much earlier. Initially, two queries are conducted, one with key 11111 (Line 5, Algorithm 4.5) and one with key 00000 (Line 2, Algorithm 4.5), which will return two entries from NL, namely $k_{first} = id_0 = 00000$ (Line 4, Algorithm 4.5) and $k_{last} = id_5 = 11000$ (Line 7, Algorithm 4.5), respectively. Hence, it can be deduced that there are no keys in $I(id_5, id_0)$. Then, as described in the following and as depicted on the right of Fig. 4.1, five iterations of the loop are executed as follows:

(1) The pair of keys $k_1 = id_0 = 00000$, and $k_2 = id_5 = 11000$ is retrieved from R. They do not share a common prefix, so spoofing with $s_1 = 01111$ and $s_2 = 10000$ discovers $x = id_3 = 01100$ and $y = id_4 = 10010$. The pairs (id_0, id_3) and (id_4, id_5) are added to the set R. After this step, it can be guaranteed that NL_{ex} does not contain keys in $I(id_3, id_4)$ since they would have been returned when spoofing IDs s_1 or s_2.

(2) The pair $(id_4, id_5) = (10010, 11000)$ is retrieved, sharing common prefix 1. The spoofed keys are thus $s_1 = 10111$ and $s_2 = 11000$. Because s_2 is identical to id_5 and hence there are no keys in $I(s_2, id_5)$, it is not necessary to spoof with s_2. Spoofing with s_1 does not result in any closer key to s_1 than id_4. No new pairs are added to R, and it is guaranteed that NL_{ex} does not contain keys in $I(id_4, id_5)$.

(3) The pair $(id_0, id_3) = (00000, 01100)$ is retrieved. Spoofing with $s_1 = 00111$ and $s_2 = 01000$ leads to the discovery of $id_1 = 00100$ and $id_2 = 01010$. Therefore, the pairs (id_0, id_1) and (id_2, id_3) are added to R. As a consequence, it is known that NL_{ex} does not contain keys in $I(id_1, id_2)$.

(4) The pair $(id_2, id_3) = (01010, 01100)$ is retrieved, but spoofing with $s_1 = 01011$ (spoofing $s_2 = 01100$ not required) reveals that NL_{ex} does not contain keys in $I(id_2, id_3)$.

(5) The pair $(id_0, id_1) = (00000, 00100)$ is retrieved, but spoofing with $s_1 = 00011$ (spoofing $s_2 = 00100$ not required) reveals that NL_{ex} does not contain keys in $I(id_0, id_1)$.

The example indicates that in each step, Algorithm 4.5 discovers a pair of keys x and y, such that it is guaranteed that the NL NL does not contain keys in $I(x, y)$. A detailed analysis to show that the observation holds for all steps and the complexity of Algorithm 4.5 is presented in [2] for interested readers.

4.1.2 Less Invasive Crawling Algorithm (LICA)

Crawling introduces communication overhead that is easily observable and may disclose the crawler to the botmasters. Although a crawler can request NLs from all known nodes iteratively, this is neither stealthy nor efficient (cf. Sect. 2.4). Current BFS and DFS-based crawling algorithms exhaustively crawl all possible nodes to provide a snapshot. The unnecessary activity of frequently requesting NLs is not only suspicions but also introduce bias if the crawl is not carried out quickly [5]. This is especially true if the reason for crawling is only to perform bot enumerations, i.e., identifying infected machines, and not to discover the full interconnectivity between bots. In the case of the former, it is desirable to minimize the necessary amount of interactions between crawler and the botnet.

The idea behind is to crawl only a subset of all nodes to obtain a *minimum vertex cover*, which is a problem known from graph theory. A *vertex cover* is a set of vertices of a graph that has all edges in the graph incident to at least one vertex of the set. The *minimum vertex cover* in a botnet is then defined as a set of minimum nodes, V_{min}, that has all other routable nodes in the network reachable from one or more nodes in the set according to our formal botnet model (cf. Sect. 2.2) as described in Eq. 4.2.

$$V_{min} = \arg\min\{|V'| \mid V' \subseteq V : \left(\bigcup_{v \in V'} NL_v\right) = V\} \qquad (4.2)$$

However, this is a $\mathcal{NP}$-hard problem and all known approximation algorithms require a global view of the graph, e.g., the algorithm of Bar-Yehuda [6].

Therefore, this work *approximates* the minimum vertex cover during a crawl. For that, this work tries to identify the *stable core* of a botnet and crawl those bots first. The MM of bots ensures information of reliable and responsive bots are regularly exchanged among all bots [5]. To exploit this observation, an iterative crawling algorithm named Less Invasive Crawling Algorithm (LICA) is proposed that employs a heuristic to plan the next crawling steps iteratively using these reliable bots and to establish a *vertex cover* in the botnet that operates with such sparse graph information. This crawling algorithm attempts to optimize the coverage of subsequent crawling steps and thus decreases the required overall number of steps for crawling a botnet. Hence, this algorithm intends not to discover the full botnet interconnectivity, but to extend the monitoring coverage to have the best, i.e., largest, snapshot of a botnet overlay along with the superpeers in it. It is assumed that bots in the botnet are all online at the same time; hence, diurnal patterns and churn effects are ignored in this work.

Algorithm 4.6: LICA($seedpeer$, R, w, t)

```
   // Initialization
 1 V_known ← seedpeer
 2 c_crawl ← 0
   // Maximum allowed requests (per node)
 3 for i = 0, i < r, i = i + 1 do
      // Utilize previous crawl
 4    | V_crawl ← V_known
 5    | V_visited ← ∅
 6    | do
      |    // Reset gain if necessary
 7    |    if c_crawl mod w = 0 then
 8    |    |  gain ← 0
      |    // select the next node to crawl
 9    |    Choose u ∈ arg max_{∀v∈V_crawl} Σ_{∀y∈V_known} |NL_y| − |NL_y − v|
      |    // Crawl + get neighborlist of u
10    |    NL_u ← crawl(u)
      |    // Update list of visited nodes
11    |    V_visited ← V_visited ∪ {u}
      |    // Update list of nodes to crawl
12    |    V_crawl ← (V_crawl ∪ NL_u) − V_visited
13    |    c_crawl ← c_crawl + 1
      |    // Calculate gain
14    |    gain ← gain + |NL_u| − |V_known ∩ NL_u|
      |    // Update visited nodes
15    |    V_known ← V_known ∪ NL_u
16    | while V_crawl ≠ ∅ & (c_crawl mod w ≠ 0 || gain ÷ w > t);
```

LICA which is described in Algorithm 4.6, not only aims at crawling efficiency but is also configurable for an adaptation to a specific environment or a specific botnet via parameters *seedpeer*, r, w, and t.

The *seedpeer* is the start node of the crawl. Parameter r is the maximum number of requests allowed to be sent, i.e., subsequent crawling iterations, to any node in the network within a particular full crawl. A full crawl ends when all nodes have been discovered.

The window parameter w determines the number of subsequent requests, for which a *gain*, e.g., ≥ 0, is calculated. The *gain* measures the number of new nodes discovered during a crawling window w. Thus, the *gain* divided by w requests gives the *learning curve* during the crawl which terminates the algorithm execution when dropping below a threshold $t, \geq 0$.

LICA utilizes the initial *seedpeer* for bootstrapping itself into the botnet overlay. Then, starting with the *seedpeer*, the algorithm obtains the NL NL_u from u (line 10) and extends its knowledge by iteratively requesting NLs from the discovered peers. For each request sent by LICA, the counter c_{crawl} is incremented by one.

Upon receiving an NL from u, it is immediately added to $V_{visited}$ (line 11) and the undiscovered peers in the received entries are added to V_{crawl} (line 15) as potential candidates for the crawl.

Line 9 in the algorithm selects the next candidate for the crawl. The algorithm goes through all received NLs of peers in V_{known} and ranks all remaining peers in V_{crawl} based on their popularity, i.e., their in-degree. Since the exact in-degree is not known, an approximation is obtained by counting the number of the occurrences a candidate or peer is seen among the NLs of other crawled peers. The function arg max returns the most-popular peer, i.e., highest ranked, as the next candidate to be crawled. In the event of equally ranked peers, the algorithm randomly chooses one among them.

At every window interval, i.e., after w requests, the algorithm checks the accumulated gain (line 14) within the past window and terminates the current crawl iteration if the ratio of the observed gain drops below threshold t (line 16). Depending on the value of r, LICA may repeat another crawl; however, this time, LICA utilizes the information of previously crawled peers V_{known} instead of the *seedpeer*. The algorithm terminates when there are no more undiscovered peers, the number of maximum allowed iterations is exceeded, or gain is below t.

4.2 Advanced Anti-crawling Countermeasures

ZEUSMILKER and LICA as presented in Sect. 4.1 circumvents existing anti-crawling mechanisms. However, it is expected that botmasters will introduce newer countermeasures to send the defenders back at trying to circumvent them yet again. Therefore in this section, by assuming the role of a botmaster, some advanced anti-crawling countermeasures that can be expected in the near future are proposed. Although mechanisms presented here can be utilized to fortify or improve future botnets, the introduction of these advanced countermeasures is necessary to trigger researchers to anticipate and be prepared before such mechanisms are seen in the wild.

In the following, two advanced anti-crawling countermeasures are proposed
to undermine some of the non-functional requirements proposed in Sect. 2.1.2.
First, improved NL restriction mechanisms over the GameOver Zeus' mechanism
(cf. Sect. 3.1.2.1) as a crawling prevention mechanism are presented. Second, a
lightweight crawler detection mechanism that is easily deployable in existing P2P
botnets is proposed.

4.2.1 Enhancing GameOver Zeus' NL Restriction Mechanism

The NL restriction mechanism of GameOver Zeus had one major weakness: the
requester can manipulate the returned entries. The existing mechanism accepts any
input, i.e., key, in the sender's message (cf. Sect. 3.1.2.1) without validating the
input or key. This *feature* allowed ZEUSMILKER to manipulate the mechanism by
deterministically spoofing keys to retrieve the bot's entire NL. Hence, it is important
that any future mechanism that aims to *prevent* crawling activities (cf. Sect. 2.4.3.1)
ensures that the requester cannot manipulate the selection of returned neighbors.
Therefore, a crawler can no longer obtain an accurate snapshot (Non-Functional
Requirement #4).

Take note that one important aspect of an MM is to ensure a robust botnet overlay.
Therefore, all NL restriction mechanisms also need to ensure that the botnet's overlay
connectivity is not adversely affected by the restriction techniques. In the following,
two countermeasures to improve the existing *GameOver Zeus* NL restriction mechanism are presented along with another countermeasure that returns random nodes
when being requested. The evaluation results of the countermeasures along with a
discussion is presented in Sect. 4.3.1.5.

4.2.1.1 Random Node Return

In Sality [7], bots return exactly one entry that is randomly chosen from their respective NLs to the requesting bot (cf. Sect. 3.2.2). Hence, the requesting bot has no
influence on the returned entries at all. Therefore, all entries have an equal likelihood to be returned. This approach from Sality is considered as one of the potential
countermeasures in this work.

4.2.1.2 Bit-XOR+

Bit-XOR+ adds additional randomness at the side of the recipient of an NL request.
The recipient, i.e., bot, generates a random key uniformly for each IP address it
receives a request from and stores it. This key is then *XOR*-ed with the key of the

requesting node, and the resulting key is then used as an input for Algorithm 4.1 to return the neighbor entries. Hence, the set of keys that is returned is now biased towards the new *XOR*-ed key, and an attacker loses its ability to strategically spoof keys. By including a randomly generated key into the selection process, each entry in the NL has the equal likelihood to be returned, such that *Bit-XOR+* is expected not to affect the connectivity between the bots negatively.

4.2.1.3 Bit-AND

Bit-AND is a variation of the *Bit-XOR+* countermeasure that executes a bit-wise *AND* operation between the stored key and the requester's key before using the resulting key to return NL entries. However, due to the nature of the *AND* operation whereupon each bit of the resulting key has a tendency to be 0 with a probability $3/4$, the set of returned keys for *Bit-AND* is likely to be biased towards keys starting with 0s. On the one hand, such a bias can considerably decrease the performance of all crawlers because the returned sets are expected to have a larger overlap in contrast to uniformly selected sets. On the other hand, keys starting with 1s are expected to be present in fewer NLs, potentially damaging the connectivity and thus the resilience of the botnet. Therefore, while *Bit-AND* is expected to achieve the best performance out of the three countermeasures, its disadvantages likely outweigh its benefits.

4.2.2 BoobyTrap: Detecting Persistent Crawlers

This section focuses on *detecting* crawlers that may still be able to tolerate the proposed NL restriction mechanisms in Sect. 4.2.1 from the perspective of a botmaster. From the observations of crawlers deployed in existing P2P botnets [8], most crawlers exhibit similar characteristics, i.e., greedy and aggressive in contacting all bots, compared to regular bots. Moreover, since researchers manually re-implement the botnet's protocol, it is also observed that many crawlers have incomplete or simplified (re)implementations of the botnet's protocol. Such characteristics of a crawler can be leveraged to detect them when crawling.

For that, Sect. 4.2.2.1 introduces a set of lightweight detection techniques called BoobyTrap (BT), which identify crawlers exhibiting peculiar characteristics. Sections 4.2.2.2 and 4.2.2.3 presents the adaptation of BT to Sality and ZeroAccess. Although similar adaptations of BT is possible for GameOver Zeus, it was not possible to evaluate it as GameOver Zeus has been sinkholed since 2014.

4.2.2.1 BoobyTrap (BT)

An BT node can be a regular bot or a sensor node (cf. Sect. 2.3.3) that is enriched with detection mechanisms or '*traps*' to autonomously identify misbehaving nodes that are contacting it. As a proof-of-concept, a sensor is implemented as a BT node through the remainder part of this work. All communication with the BT node is logged with corresponding metadata: timestamp, payload, source IP, and source port. Compared to conventional sensors, BT nodes can have additional functionalities such as responding to NL requests with valid replies and (re)sending valid probe messages to the sender of a request message. BT nodes can also listen for incoming connections or messages on a secondary port (if required), as it is needed for one of the traps (described later). Take note that in this work, the deployed BT nodes only return non-bot entries, i.e., other sensor nodes, to avoid participating in the regular botnet maintenance activities due to legal constraints.

The BT mechanism leverages upon the following assumptions of crawlers that are derived from the observations in real-world botnets.

1. Crawlers greedily attempt to discover/contact all bots by aggressively abusing the botnet's protocol to request NL.
2. Crawlers cannot distinguish BT nodes from normal bot without first interacting with them.

The main idea of BT is to identify "misbehaving" nodes, i.e., crawlers, by distinguishing their behavior from bots on the basis of violations of the respective botnet's *MM* protocol. These behaviors can be categorized according to the following classes: *Defiance*, *Abuse*, and *Avoidance*, that are generic to any P2P botnet.

Defiance

Bots that defy the botnet-specific MM protocols can be classified as crawlers. An example of such defiance includes omitting certain prerequisite actions or mandatory message exchange(s) before requesting an NL. In some cases, it is also possible to identify a crawler based on its behavior of contacting *all* discovered entries, even when the botnet protocol applies certain restriction to entries that should be chosen as potential neighbors. For example, entries that have a matching /20 subnet with an existing entry are ignored by GameOver Zeus (cf. Sect. 3.1.2.2). As such, if a *BT* node returns an entry of another (BT) node that is from the same /20 subnet, and if new node is contacted by the same node, this behavior can be classified as a crawler.

Abuse

In P2P botnets, the ability to request new neighbors is necessary to prevent getting isolated from the botnet overlay. However, crawlers make use of this NL exchange mechanism to reconstruct the network topology of the botnet (cf. Sect. 2.3.2). Therefore, bots in most recent P2P botnets return only a subset of their NL to prevent a crawler from retrieving the entire NL easily (cf. Sect. 2.4.3.1). Since the presence of churn also encourages crawlers to obtain snapshots of the botnet as fast as possible to avoid introducing bias in the monitoring results (cf. Sect. 2.4.1), crawlers typically crawl aggressively [2]. In contrast, bots usually probe their neighbors only once per the defined interval of the MM-cycle. Thus, a frequency-based detection mechanism can be utilized to detect crawlers that abuse the NL exchange mechanism. In fact, such a countermeasure is already implemented by the *GameOver Zeus* botnet (cf. Sect. 3.1.3).

Avoidance

The MM of a botnet defines the sequence of message exchange as well as the structure of the messages required for botnet communication. Crawlers may omit some of it, which could be optional, to reduce communication overhead or to refrain from helping the botnet, e.g., sharing of botmaster commands. By deliberately sending botnet-specific requests that would otherwise generate a verifiable reply, crawlers that are refusing to respond (or ignore the requests) can be detected accordingly.

4.2.2.2 Adaptation of BoobyTrap for Sality

In the following, three crawler detection mechanisms or *traps* are adapted for Sality, adhering to two out of the three misbehavior classes presented in Sect. 4.2.2. Each trap's name for Sality is prefixed with an *S*, and the abbreviation of the respective detection class. There are two traps for Sality in the class of *Defiance*, i.e., *SD1-IgnoreTrap* and *SD2-BaitTrap*, and one trap from the class of *Abuse*, i.e., *SAB-BurstTrap*. Take note that a trap can also be set up from the class of *Avoidance* based on the appended *URLPack* within the *Hello* message exchange process. However, such a trap for Sality is omitted in this work as it may induce self-DDoS on the BT node itself as an amplification attack [9].

SD1-IgnoreTrap

The MM protocol of Sality dictates that a bot uses a *Hello* message to probe the responsiveness of its neighbors (cf. Sect. 3.2.2). If the neighbors are responsive, and if the probing bot requires additional neighbors, only then, it sends an additional NL_{Req} message. As such, an NL_{Req} is *always* preceded by a *Hello* message. Crawlers, that want to simplify this process to reduce the communication overhead for crawling, may decide to ignore the *Hello* message and send only NL_{Req} messages to the bots. Moreover, simplifying the process also reduces the amount of time required for the crawler to obtain a snapshot. For each received NL request, the BT node checks if there has been a preceding *Hello* message logged in the database. If no record exists for the *Hello* message, the node is flagged as a crawler.

SD2-BaitTrap

The MM protocol of Sality also ensures that an IP address can only be present once in a bot's NL (Line 3, Algorithm 4.2). When a bot discovers a potential neighbor with the same IP but different port (Line 5, Algorithm 4.2), it omits the new entry. This trap exploits this behavior by deliberately responding to all received NL_{Req} with an entry that points back to the BT node's secondary port. Legitimate bots would ignore such a reply, since the initial entry, i.e., the entry with the primary port, is still responsive in the NL from the previous MM cycle. Since crawlers are often greedy, the crawlers may also probe the secondary port of BT and trigger the detection mechanism. Take note that this particular *baiting* mechanism can also be executed using two (or more) colluding BT nodes.

SAB-BurstTrap

Bots in Sality probe the responsiveness of their neighbors once every 40 min (cf. Sect. 3.2.2). Bots can also (optionally) request neighbors of their neighbor by sending an NL_{Req}. As such, this trap keeps track of a bot's NL requesting frequency, i.e., based on the IP address of the requester. If a bot sends too many NL requests within a short interval, i.e., < 40 min, this behavior triggers the detection mechanism.

4.2.2.3   Adaptation of BoobyTrap for ZeroAccess

In the following, three crawler detection mechanisms or *traps* are adapted for ZeroAccess from the different misbehavior classes presented in Sect. 4.2.2. Each trap's name is prefixed by an abbreviation of the botnet's name and the respective detection class.

ZD-NonComplianceTrap

The MM protocol of ZeroAccess allows bots to identify if a *getL* message was received. This is done by checking the flag value of the received request message, i.e., $flag == 0$. In reply, legitimate bots *always* send a *getL+* message that has its flag set to 1. However, it would still be protocol-compliant if a bot sends a *getL+* message with the flag set to any non-zero integers, i.e., $flag \neq 0$. This trap deliberately sends a *getL+* with a modified flag-value, e.g., $flag = 3$, for every received *getL* message. A legitimate bot will respond to all requests with *retL* or *retL+* messages that have the *flag* values *copied* from the received request messages (Line 2). In addition, due to the possibility of UDP hole-punching in ZeroAccess [1], all legitimate bots (including those behind NAT-like devices) should respond to any *getL+* message received. Hence, the BT node examines whether the received replies contain inconsistent flags and detect the crawlers accordingly.

ZAB-BurstTrap

The MM mechanism of ZeroAccess indicates that a bot would only contact a particular neighbor at most three times within a duration of 256 s (cf. Sect. 3.3.2). This observation is exploited in this trap that triggers when any bot attempts to contact the BT node aggressively in quick successions, i.e., more than three requests within 256 s. Similar to *SAB-BurstTrap*, if more than three requests are received from a single bot within a short interval, i.e., < 256 s, this detection mechanism is triggered.

ZAV-IgnoreTrap

This trap works in tandem with the *ZD-NonComplianceTrap*. Crawlers that received the BT node's *getL+* requests with modified flag values may (intentionally or unintentionally) decide not to respond to the message. Considering the fact that UDP hole punching is exploited, all bots including those behind NAT are expected to respond to all valid requests. Therefore, any node deliberately refuse to reply can be flagged as a crawler.

4.3   Evaluation

This section presents the evaluation results and analysis of the mechanisms proposed in Sects. 4.1 and 4.2 in three parts. In the first part (Sect. 4.3.1), a thorough analysis of ZEUSMILKER is presented in the context of circumventing the NL restriction mechanism of GameOver Zeus as described in Sect. 4.1.1. In addition, the evaluation of the enhanced restriction mechanisms as introduced in Sect. 4.2.1 is also presented.

The second part of this section (Sect. 4.3.2) presents the evaluation results of the Less Invasive Crawling Algorithm (LICA) as described in Sect. 4.1.2. The final part (Sect. 4.3.3) provides an analysis of the ability to detect crawlers in existing botnets through the mechanisms introduced in Sect. 4.2.2.

4.3.1   Evaluation of ZEUSMILKER

The evaluation of ZEUSMILKER is outlined below. First, the dataset utilized for evaluating ZEUSMILKER is discussed in Sect. 4.3.1.1. Then, Sect. 4.3.1.2 elaborates on the setup for the experiments and Sect. 4.3.1.3 introduces the metrics used in the evaluations. After that, the investigated research questions are listed along with the expectations of the outcome in Sect. 4.3.1.4. Finally, the results of the experiments are presented in Sect. 4.3.1.5.

4.3.1.1   Dataset

A real-world GameOver Zeus dataset is used in the evaluation that consists of crawl information collected for a duration of approximately five hours from the botnet on 25th April 2013. The sanitized dataset contains information of 900 bots that have between 10 to 70 entries in their respective NL. The median of the dataset is 34 entries with a standard deviation of 18.37.

4.3.1.2   Experimental Setup

The MM of GameOver Zeus was implemented in *OMNeT++*[1] by making use of *OverSim*[2] [10] as the simulation framework. OMNeT++ is a discrete event simulator that allows simulation of networks. OverSim adds the required functionality for overlays that is leveraged in implementing the membership management mechanism of the botnets. For ZEUSMILKER, the implementation includes the NL restriction mechanism as described in Algorithm 4.1. As for the generation of random keys

---

[1] http://www.omnetpp.org

[2] http://www.oversim.org

within *OverSim*, the *OverlayKey* class is utilized to generate keys following a uniform distribution to investigate the effects of key distribution within the NLs.

For each iteration of the experiment, data for each bot in the simulation is uniformly selected at random from the dataset described in Sect. 4.3.1.1 depending on the investigated NL size. The bots are then assigned the selected key and have their NL filled with the associated NL entries. Since the order of entries in a bot's NL is non-deterministic, a random permutation is applied to the contained entries before storing them as the NL.

The following approaches were also implemented in the simulation framework to evaluate the effectiveness of the NL restriction mechanism in GameOver Zeus as well as the efficiency of ZEUSMILKER in comparison to the following approaches:

- **ZEUSMILKER** is the proposed approach for strategically spoofing keys to *milk* all entries from a bot's NL implemented as per Algorithm 4.5.
- **Random** is the only other known monitoring approach for monitoring GameOver Zeus [1]. The spoofed keys are 160-bit in length and generated uniformly at random for each request.
- **BinaryHalving** spoofs keys by halving the ID space in the manner of a binary search algorithm. For each iteration of the algorithm, two keys are derived between two previously crawled keys. This halving process is repeated until the maximum number of permitted requests is reached. For that, *BinaryHalving* initially spoofs with $\mathbf{0}(b)$ and $\mathbf{1}(b)$, and adds the pair $(\mathbf{0}(b), \mathbf{1}(b))$ to a FIFO queue Q. Then it executes the following statement T times:

  1. Remove the head (K_1, K_2) of Q and determine the keys $h_1 = \lfloor \frac{K_1 + K_2}{2} \rfloor$ and $h_2 = h_1 + 1$,
  2. Crawl using spoofed keys h_1 and h_2, and
  3. Add (K_1, h_1) and (h_2, K_2) to Q.

For each experiment, the results were averaged over 50 independent trials with confidence intervals of 95%. Furthermore, for each iteration of the experiments, a unique seed value has been used to initialize the simulation models and to choose a random node from the dataset. In all of the experiments, the maximum number of requests is limited to $2n$ requests, where $n = |NL|$.

4.3.1.3 Evaluation Metric

The success of anti-crawling countermeasures and the performance of ZEUSMILKER is measured by the *discovery ratio*. It is defined as the unique fraction of an NL that is retrieved during crawling. Hence, the discovery ratio is an assessment of both the efficiency of the crawling algorithm as well as the effectiveness of the botnet's countermeasures, allowing the comparison of different crawling and anti-monitoring strategies.

4.3.1.4   Research Questions and Expectations

One of the countermeasures to hamper successful botnet monitoring is to restrict the number of entries that are returned after receiving NL request (cf. Sect. 2.4.3.1). Hence, the following research question needs to be answered in the evaluation:

- *What is the influence on different NL sizes n and NL return sizes l?*

ZEUSMILKER is expected to obtain all entries successfully with at most $2n$ requests in every scenario. Meanwhile, *Random* and *BinaryHalving* are expected to miss some entries. The *Random* crawling strategy retrieves a randomized set of entries and has a high probability of missing a few keys. *BinaryHalving*, in contrast, divides the search space strategically, but does not make use of knowledge gained in previous steps and as such may continue to query regions with few or no keys intensively. Furthermore, the performance of all algorithms is expected to improve with increasing l, because more keys are discovered in each step. The increase should be particularly strong for *Random* as the probability to be successful when spoofing randomly is highly dependent on the number of trials.

The distribution of keys within a real world GameOver Zeus bot's NL is observed biased to the key of the bot itself [1]. However, due to the botnet's NL return mechanism (cf. Sect. 3.1.2.1), different key distributions may influence the number of requests needed to be able to retrieve the entire NL.

Examples of other distributions that could occur in a bot's NL are:
1. **Random Distribution**: A node's NL contains only randomly generated keys.
2. **Consecutive Entries**: A node's NL contains only consecutive keys, e.g., $k_{j+1} = k_j + 1 \mod 2^{160}$.

Therefore, the following research question needs to be answered in the evaluation:

- *How do the different distributions of keys within a bot's NL influence the performance of the spoofing algorithms?*

ZEUSMILKER is expected to retrieve all entries within a bot's NL independent of the chosen distribution. However, in the case of *Consecutive Entries*, it is expected to retrieve all unique entries in only n requests instead of $2n$, as it is not necessary to check for additional keys between two neighboring keys. In contrast, *Random* and *BinaryHalving* are expected to require more crawling requests in this setting. Especially, *BinaryHalving* is expected to perform worst, as it spoofs many keys that yield no new knowledge in the *Consecutive Entries* setting. However, in the *Random Distribution* setting, both are expected to be closer, but still inferior to the crawling performance of ZEUSMILKER, as a result of the uniform key distribution.

The existing NL restriction mechanism of GameOver Zeus is exploitable as the requesting bot can manipulate the choice of returned entries based on the supplied key. Newer countermeasures presented in Sect. 4.2.1 attempt to either deny the possibility of manipulating the entries at all or allow restricted degree of manipulation on the returned entries. As such, the following research question needs to be answered in the evaluation:

- *How does the crawling algorithms perform in the presence of countermeasures like Random Node Return, Bit-XOR+, and Bit-AND?*

For *Random Node Return*, both *Random* and *BinaryHalving* algorithms are expected to be comparable in performance since the returned entries are not influenced by their key spoofing algorithm. However, *ZeusMilker* is expected to perform poorly if the choice of keys returned by the bot mislead the algorithm into believing that there are no more keys left undiscovered.

For *Bit-XOR+*, *Random* is expected to perform best compared to the other two algorithms as it produces higher entropy of keys used in selecting neighbors to be returned. Finally, for *Bit-AND*, all algorithms are expected to perform poorly since the bits of the returned entries are highly biased to 0 with a probability of $3/4$.

4.3.1.5   Results

In the following, the findings are summarized on the impact of parameters l, n, the assumed key distribution, and effectiveness against advanced countermeasures on the three different crawling algorithms.

Impact of the size of the returned neighborlists l

First, the impact of the size of the returned NL l is discussed. Figure 4.2a summarizes the discovery ratio for a default parameter setting of GameOver Zeus with $n = 50$ and $l = 10$ in dependence on the number of requests for all three crawling strategies. As can be observed, ZEUSMILKER can successfully retrieve all entries in a bot's NL within 100 requests. At the same time, *Random* discovers only 92% and *BinaryHalving* only 53% of all entries in a bot's NL. Thus, the results confirm the expectation that *BinaryHalving* is not suitable for such biased NLs. *BinaryHalving* performs poorly because of retrieving many duplicate entries as a result of spoofing keys within a range of the key space that provides no additional new information. For all algorithms, the number of initially retrieved entries increases fast with only a few queries. Later on, when only a few keys are left undiscovered, the slope of the performance curve decreases. Note that during the first few queries, the *Random* crawling algorithm even manages to discover more number of unique keys than ZEUSMILKER. A potential reason for the initially weaker performance of ZEUSMILKER is the choice of the two spoofed keys s_1, s_2 (see Eq. 4.1), which are potentially very close and hence can result in returned sets with a high overlap. However, ZEUSMILKER is clearly superior to *Random* and *BinaryHalving* in discovering larger portions or even the complete NL.

Figure 4.2b shows the discovery ratio in dependence on the number of crawling requests for $n = 50$ and $l = 1$. ZEUSMILKER still retrieves all entries within the predicted 100 requests, though slower than for $l = 10$. As only one entry per request can

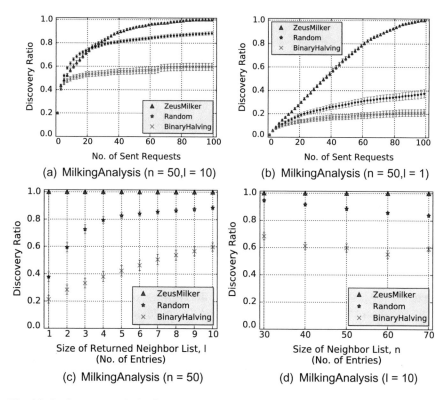

Fig. 4.2 Performance analysis of ZEUSMILKER, *Random*, and *BinaryHalving* on GameOver Zeus for various NL sizes n and returned NL sizes l. *Source*: https://ieeexplore.ieee.org/stamp/stamp. jsp?tp=&arnumber=7164947

be obtained, the number of retrieved keys initially increases linearly and then converges slowly to a discovery ratio of 1. The decrease in performance is more apparent for *Random* and *BinaryHalving*: The discovery ratio for both approaches decreases drastically compared to $l = 10$, to 19% for *BinaryHalving* and 37% for *Random* at 100 requests. A more detailed analysis of the impact of l is given in Fig. 4.2c showing the discovery ratio in dependence on l for $n = 50$. ZEUSMILKER successfully obtains all neighbor entries within $2n = 100$ queries independent of l, whereas the discovery ratios after $2n$ queries of the other two strategies are significantly affected by l. Since both algorithms are unable to strategically spoof keys, the fraction of retrieved keys drastically decreases when the number of returned keys l is reduced. Hence, the results of this analysis match the initial expectation that smaller values of l restrict the amount of new knowledge the crawling algorithms could obtain. However, since ZEUSMILKER can strategically spoof keys to discover all entries in a NL, its ability to retrieve the complete list remains unaffected by different values of l.

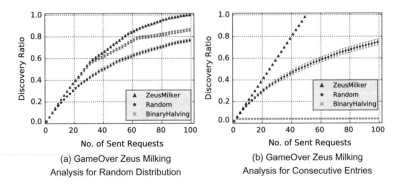

Fig. 4.3 Performance analysis of ZEUSMILKER, *Random*, and *BinaryHalving* for different key distributions in NLs (**n = 50, l = 1**). *Source*: https://ieeexplore.ieee.org/stamp/stamp.jsp?tp=& arnumber=7164947

Impact of the size of the neighborlists n

Next, the impact of the size of the NL n on the crawling performance is analyzed. Figure 4.2d shows the discovery ratio of the different algorithms in dependence on n for $l = 10$. Independent of n, ZEUSMILKER successfully discovers all nodes in an NL. In contrast, the performance of *Random* slowly decreases with increasing n, because it is harder to discover large sets simply by random trials than on smaller sets. The slight decrease in performance of *BinaryHalving* is not significant.

Influence of different key distributions

Apart from n and l, different key distributions in NLs may also influence the performance of crawling algorithms. Figure 4.3a shows the discovery ratio in dependence on the number of requests for all three crawling strategies in the *Random Distribution* setting. As expected, ZEUSMILKER can obtain all entries with at most $2n = 100$ requests whereas *Random* and *BinaryHalving* only discover about 80 and 90% of all NL entries, respectively. Both strategies perform considerably better than the results for the real-world data set, increasing their discovery ratio by more than a factor of 2 and 4, respectively. This improved performance is contributed by the well-distributed keys, resulting in fewer duplicates during successive crawling attempts. As expected, *BinaryHalving* also performs much better than *Random* when the uniform key distribution assumed by *BinaryHalving* is indeed given.

The inability of *BinaryHalving* to deal with non-uniform key distributions becomes evident when considering *Consecutive Entries*. Most of the time, *BinaryHalving* discovers only two keys, i.e., discovery ratio is about 4%. A potential reason is the repeated spoofing of keys at distances away from all keys in the NL. As a result, the same two keys are returned repetitively. ZEUSMILKER, in contrast, can successfully discover all entries with only n requests instead of $2n$ because the sets $I(K_j, K_{j+1})$ are empty so that no additional n keys need to be spoofed to verify that $I(K_j, K_{j+1}) \cap NL = \emptyset$. In contrast, the performance of the *Random* crawling is similar to its performance when considering randomly distributed keys.

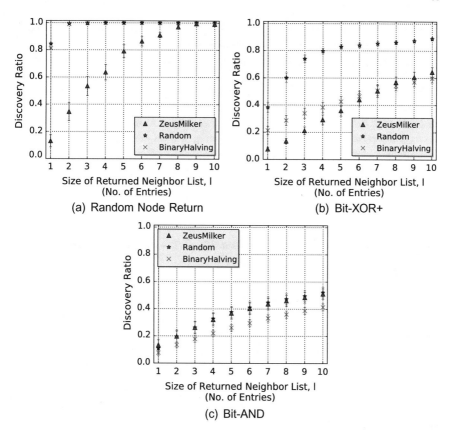

Fig. 4.4 Performance analysis of ZEUSMILKER, *Random*, and *BinaryHalving* on the presence of different advanced countermeasures (**n = 50**). *Source*: https://ieeexplore.ieee.org/stamp/stamp.jsp?tp=&arnumber=7164947

Enhanced Restriction Mechanisms

The evaluation results of the proposed anti-crawling countermeasures in Sect. 4.2.1 as depicted in Fig. 4.4 is presented in the following. The *Random Node Return* analysis in Fig. 4.4a indicates the inefficiency of this countermeasure in restricting the crawler. Both *Random* and *BinaryHalving* were able to retrieve more than 80% of a bot's NL in all parameter settings after 100 requests. However, ZEUSMILKER performed poorly due to the inherent incorrect assumptions made on the returned keys that led the algorithm to falsely assume there are no more keys left undiscovered.

The results for the *Bit-XOR+* countermeasure indicate that *Random* performs best with an average of about 80% of nodes discovered for $l \geq 4$, as displayed in Fig. 4.4b. Hence, the performance of *Random* is largely not influenced by bits flipping, as can be seen from comparing Fig. 4.4b and Fig. 4.2c, depicting the performance of the crawling for the unaltered GameOver Zeus. Although *BinaryHalving* initially performs better than ZEUSMILKER for $l \leq 7$, its performance degrades for $l > 7$, as a result of spoofing keys that yield more duplicate entries. However, ZEUSMILKER's strategy of deriving keys based on previous knowledge provides more randomness, i.e., a variety of key prefixes, in the spoofed keys, hence obtains a slight improvement than *BinaryHalving* towards the end.

Bit-AND, as displayed in Fig. 4.4c, presents a better restriction mechanism than *Random Node Return* and *Bit-XOR+* as the discovery ratio of all crawling algorithms is kept below 50% for $l \leq 10$. The discovery ratio increases with the size of the returned NLs l in a close to linearly manner. Although the poor performance of all strategies in terms of discovery ratio indicates the effectiveness of this countermeasure, the bias resulting from this strategy may negatively affect the robustness of the resulting overlay.

4.3.2   Evaluation of the Less Invasive Crawling Algorithm (LICA)

The evaluation of LICA is outlined below. First, the dataset utilized for evaluating LICA is discussed in Sect. 4.3.2.1. Then, Sect. 4.3.2.2 elaborates the setup for the experiments and Sect. 4.3.2.3 introduces the evaluation metric. After that, the investigated research questions are listed along with the expectations of the outcome in Sect. 4.3.2.4. Finally, the results of the experiments are presented in Sect. 4.3.2.5.

4.3.2.1   Dataset

Two different real-world unstructured P2P network datasets in the form of directed graphs, i.e., GameOver Zeus and *Gnutella*, were used to evaluate the performance of crawling algorithms.

The GameOver Zeus dataset used in this evaluation consists of crawling information collected in approximately five hours from the GameOver Zeus botnet on 25th April 2013. It has been obtained from previous work in analyzing GameOver Zeus [1]. From the initial $1,061,402$ edge entries in the database, $667,704$ edge entries were removed that consist of biases that were made known by the authors: sinkholed nodes (identified by an out-degree < 10), sensor nodes (identified by an in-degree > 500), and duplicated edges. Take note that, since the crawl data consists of multiple continuous crawls over a longer period, some bots may have reported a lot of neighbors than usual, i.e., 50. This is mainly due to churn dynamics within the botnet (cf. Sect. 2.4.1).

The second dataset is the crawl data of the unstructured P2P file-sharing network *Gnutella*, in August 2002 that was obtained from the SNAP repository.[3] This dataset was used as it is for the evaluation, i.e., without any sanitization. A summary of

---

Table 4.1 Graph properties of the datasets. *Source*: https://ieeexplore.ieee.org/stamp/stamp.jsp? tp=&arnumber=6883429

Dataset name	Gameover zeus	Gnutella
Nodes	82,471	62,586
Nodes (out-degree > 10)	10,794	16,387
Avg. NL size	4.6	2.4
Highest NL size	97	78
Edges	379,088	147,892
Avg. clustering coefficient	0.01934	0.00047
Diameter	11	31
Avg. path length	5.2	9.2

the dataset properties is provided in Table 4.1. Please note that properties listed for GameOver Zeus in the table are the post-sanitize properties of the dataset.

4.3.2.2   Experimental Setup

The analysis was conducted using *Python* and the *NetworkX* module [11], with all crawling algorithms discussed in Sect. 4.1.2, i.e., Less Invasive Crawling Algorithm (LICA), Breadth-First Search (BFS), and Depth-First Search (DFS), implemented as Python scripts. To address the issue of some nodes having NL size of more than 50, all entries of a bot are shuffled and split into a sequence of chunks, i.e., 50 entries in each chunk. This value is chosen to resemble closely the GameOver Zeus' implementation of the NL size.

For every received NL request from the implemented crawling algorithms, a node will return a single chunk from its sequence and repeats the sequence after returning the last chunk (if queried further). In the experiments, a full crawl ends when there are no more new peers to crawl. In addition, LICA also ends its crawl when the maximum allowed iterations have exceeded (cf. Sect. 4.1.2).

Fifty (50) independent experiments were executed on each of the experiments and the final results were averaged over them. For each iteration of the experiment, the simulation uniformly chooses a common seed peer to begin crawling for all the algorithms. Furthermore, for the clarity of the resulting plots, all algorithms terminate their crawling as soon as 95% (indicated by the horizontal dashed lines) of nodes in the datasets have been discovered unless stated otherwise.

4.3.2.3   Evaluation Metric

To evaluate the performance of a crawling algorithm, a *ratio of discovered peers* is used as an evaluation metric. This metric represents the ratio of *unique* peers discovered dependent on the number of sent request messages.

Meanwhile, to evaluate the efficiency of a crawling algorithm, the local-ratio approximation of *minimum vertex cover* presented by Bar-Yehuda et al. [6] is utilized. This approximation provides a value of the number of minimum nodes to be crawled to obtain a snapshot of the botnet graph, i.e., a vertex cover. The approximation ratio of this algorithm is $2 - \frac{1}{k}$, where k is the smallest integer. The implementation of this approximation algorithm that is available in *NetworkX* operates on undirected graphs. Hence, it has been modified to be applied to directed graphs. In the remainder of this evaluation, this modified algorithm is referred as *Approximative Minimum Vertex Cover (AMVC)*.

4.3.2.4 Research Questions and Expectations

LICA is a flexible crawling algorithm that requires a combination of parameters to perform well. Depending on the choice of the parameters, an ongoing LICA-based crawl may terminate quickly or much later. However, the process of selecting the best combination of parameters w, r, and t is not very intuitive. For that, the following research question needs to be answered:

- *What is the best combination of parameters for LICA in GameOver Zeus and the Gnutella dataset?*

DFS and BFS-based crawling algorithms attempt to discover all bots in a greedy manner. However, not much have been discussed on the performance of these algorithms in a real world botnet scenarios. Therefore, the following research question needs to be answered:

- *How well do the different crawling algorithms perform on real world datasets?*

LICA is expected to perform better compared to the other algorithms as it prioritizes backbone nodes and terminates as soon as the ratio of discovery falls below the threshold value. BFS is expected to perform better than DFS due to candidate prioritizing strategy adopted by this algorithm, i.e., first come first serve, which discovers more number of new peers than its counterpart.

In addition, since the best results – relative to the goals of LICA – in enumerating all bots in a botnet is to obtain a *minimum vertex cover*, it is of interest to identify which crawling algorithm perform closest to the *AMVC* value. Therefore, the following research question needs to be answered:

- *Which crawling algorithm provides best efficiency?*

LICA is expected to perform the best due to the presence of a natural backbone that can be leveraged by the crawlers. Meanwhile, BFS is expected to perform also slightly better than DFS due to the candidate selection strategy that discovers more number of unique peers in the beginning.

4.3.2.5 Results

Rationale

Existing crawling algorithms which are implemented using *BFS* or *DFS* methods to crawl botnets may not be efficient. The node selection criterion used by both these algorithms is not based upon any other information except the order the nodes are stored and processed. Therefore, a series of experiments is conducted to understand the impact of the node selection criteria on the crawling performance of LICA.

Best combination of parameters for LICA

First, it is investigated on when to terminate an ongoing crawling process to avoid too many unnecessary crawling steps. For that, LICA contains a simple mechanism that checks the *gain* after a window w of crawling steps, and that terminates when the *gain* drops below threshold t during the crawl. As the algorithm utilizes a *learning curve* to terminate the crawl, the algorithm is adjustable by manipulating its parameters. For example, to overcome the blacklisting mechanism of GameOver Zeus, [4] the R value can be set to 11, i.e., maximum number of requests that are allowed to be sent to a particular node. Alternatively, subsequent full crawls can be delayed by 60 s. The value R is set to 2 in all experiments because it is known from the GameOver Zeus datasets, that all algorithms need to request the NLs from any node at most twice, i.e., two chunks, to obtain the full NL.

Also, by deciding combinations of values for the window size w and threshold t, the resulting crawl can also be shaped. For example, by specifying a high threshold value, e.g., > 1.0 in combination with a high window size, e.g., 3000, the backbone nodes may be targeted for crawling. Similarly, when the intention is to crawl as many nodes as possible, the threshold value can be set to a relatively low value, e.g., < 0.05 in combination with a low window size, e.g., 300. The values of the window size and threshold used in this work were obtained through a parameter study with various combinations. The effects of different combinations of promising values are analyzed on the GameOver Zeus dataset with the value $R = 2$ as presented in Fig. 4.5a.

From the analysis, it is identified that when the threshold value t, is low, e.g., 0.1, and with the window value w of 300, a full crawl results in about 94.7% of the entire dataset known just with about 29, 000 sent requests. Meanwhile, the parameter combination of $t = 0.45$, $w = 1000$ obtained a lower coverage of 93.3% although with 17.4% fewer requests than the previous combination. However, with an increased threshold value, e.g., 0.8, and window value $w = 2000$, LICA terminates with a coverage of 93.9% despite requiring 1, 800 requests more than the previous combination. Therefore, it is decided that the combination, $t = 0.8$, $w = 2000$ is more reasonable to crawl the GameOver Zeus dataset efficiently. Unfortunately, the criteria for selecting the best combination of parameters are not straightforward, i.e., the combination of parameters can only be selected based on the basis of a trial and

---

[4]At the time of this work, GameOver Zeus allowed 12 requests to be received within a sliding window of a minute. This value was later changed in a subsequent botnet binary update to only 6 requests. (cf. Sect. 3.1.3)

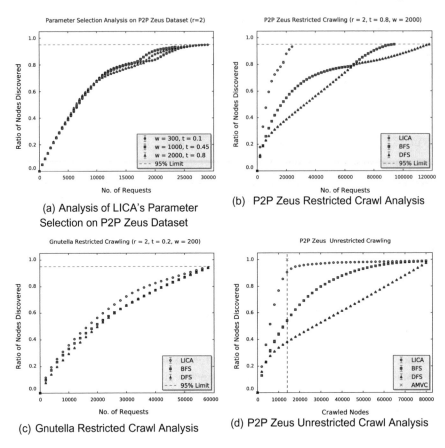

Fig. 4.5 Performance analysis of LICA, *BFS*, and *DFS*. (**a**) The performance of LICA under different combination of parameters. The performance of all observed crawling algorithms on (**b**) GameOver Zeus and on (**c**) the *Gnutella* dataset, measured in the ratio of nodes discovered in dependence on the total number of requests sent. (**d**) contains the results of all crawling algorithms on the GameOver Zeus dataset without any NL restrictions, plotted by the ratio of nodes discovered depending on the total number of nodes crawled. *Source*: https://ieeexplore.ieee.org/stamp/stamp.jsp?tp=&arnumber=6883429

error. Although some general combination of values can be utilized, based on the requirement of the crawl, LICA performs better if fine-tuned with more appropriate parameters to improve the crawling efficiency, i.e., based on results from previous crawls.

Performance analysis of crawling algorithms

The results of the crawl performance analysis are presented in Fig. 4.5b and 4.5c. LICA was executed on the GameOver Zeus dataset using the previously chosen parameter combination: $r = 2$, $t = 0.8$, and $w = 2000$. The crawl performance in Fig. 4.5b indicates a much better performance of LICA in comparison to the other methods. For LICA, the points in which there were results less than 16 individual experiments were omitted as a confidence interval cannot be obtained from that. LICA required 25, 780 requests to obtain a 93.86% coverage of the peers in the botnet. This is about 27% of the total requests made by the *BFS* algorithm to obtain a 95.0% coverage. *DFS* performed worst in this analysis by requiring additional 400% requests than needed by LICA.

The convergence point between the *BFS* and *DFS* algorithm indicates the point where all known nodes during the initial crawl have been crawled. The growth that is observed after that point is from new nodes discovered from re-requesting NLs from previously known nodes, i.e., subsequent chunks of their NL. This convergence behavior is not observed in LICA because it terminates the crawling when the observed gain drops below the threshold, and the gain immediately picks up in the subsequent crawl iteration.

It is worth mentioning that by reducing the size of the NLs or the returned subset of the list, the effort to crawl the entire network increases proportionally for all crawling algorithms. This is verified by running another set of experiment with a returned NL of size 30 and $R = 3$. The observed performance between the crawling algorithms remain relatively similar to the results in Fig. 4.5b.

The experiment is repeated on the *Gnutella* dataset using the following parameter combination: $r = 2$, $w = 400$, and $t = 0.3$. However, the performance gain of LICA for *Gnutella* dataset in Fig. 4.5c is not as significant as in the GameOver Zeus dataset. Further investigation revealed that this behavior is due to the diameter of this dataset being very high, i.e., 31, with an average path length of 9.2. Moreover, nodes in this dataset have a rather low average size of the NL, e.g., 2.4 entries. Hence, due to the inherent network structure in this dataset which is unlike the structure of most P2P botnets, the gain is much lower, as all crawlers need to go through almost every available node to obtain a full view. Nevertheless, the performance of LICA is better compared to the other two algorithms as presented in Fig. 4.5c. For example, with 31, 941 requests, LICA discovered 75.5% of nodes that is about 7% more than the other algorithms.

Efficiency analysis of crawling algorithms

For this purpose, the performance of all the three crawling algorithms on the GameOver Zeus dataset are compared with respect to the *AMVC* value in Fig. 4.5d. The simulation settings were modified to allow all available neighbors of a node to be returned in a single request and disabled the crawl termination mechanism in LICA. As such, the purpose of this particular analysis is to find out how many nodes need to be crawled to obtain the full view of the network.

Based on this analysis, it is demonstrated by heuristic that LICA outperforms other methods in performing closer to the calculated *AMVC* value, 14, 050 nodes. At the point of the *AMVC*, LICA discovered a total of 90.6% nodes in comparison with *BFS* that only discovered 54.1% or *DFS* with 38.2% of nodes. This is interesting because it indicates that by crawling and prioritizing the '*popular*' peers, the backbone of the network is being leveraged and crawled. This corresponds to the finding of Stutzbach et al. [5] that reports the existence of biased connectivity with peers with higher uptime, i.e., *popular* nodes in this work. This allowed LICA to exploit this feature and outperform existing crawling algorithms.

4.3.3 Evaluation of the BoobyTrap Mechanism

The evaluation of the BoobyTrap (BT) mechanisms as proposed in Sect. 4.2.2 is outlined next. First, the dataset utilized for evaluating BT is discussed in Sect. 4.3.3.1. Then, Sect. 4.3.3.2 elaborates the setup for the experiments. After that, the investigated research questions are listed along with the expectations of the outcome in Sect. 4.3.3.3. Finally, the results of the experiments are presented in Sect. 4.3.3.4.

4.3.3.1 Dataset

The datasets that were used for evaluation were obtained using a real deployment of BT nodes, i.e., sensors, in Sality *Version 3* (cf. Sect. 3.2) and ZeroAccess *Network 2* (port 16470) (cf. Sect. 3.3). Each BT node was popularized for two weeks before the measurements were obtained. After that, the measurements were collected for a duration of one week in each botnet: Sality (23/09/2015 00:00:00 CET to 29/09/2015 23:59:99 CET) and ZeroAccess (02/10/2015 15:57:55 CET to 09/10/2015 15:57:54). Table 4.2 presents the summary of the datasets.

4.3.3.2 Experimental Setup

The experiments were conducted with BT-enhanced sensors that were implemented in *Python* language for both botnets, i.e., Sality and ZeroAccess. All detection mechanisms were triggered by the type and contents of the responses received (or missing) from a node. However, only for the frequency-based detection mechanisms, i.e., *Abuse* class, a configurable sliding window-based detection mechanism was implemented to help identify IPs of aggressive crawlers. This detection mechanism takes two input parameters: length of the sliding window t (in seconds) and the mini-

Table 4.2 Statistics of the collected data. *Source*: https://ieeexplore.ieee.org/stamp/stamp.jsp?tp=&arnumber=7510885

	Sality (Version 3)	ZeroAccess (Port 16470)
Total IPs	735, 443	25, 236
Average IPs/day	162, 804	7, 128
Min IPs/day	155, 957	5, 905
Max IPs/day	177, 267	7, 864

mum number of messages n_{min} to trigger the detection mechanism. If a particular node from an IP address sent more than n_{min} messages within any observed sliding window, a detection will be triggered.

To evaluate the performance of the proposed mechanisms, the amount of IP addresses that triggered the BTs were considered and manual verification was conducted on the log data to identify if the behavior of a node behind an IP matched with that of a possible crawler. Some of the characteristics that were inspected and considered are listed in the following:

- Rate of consecutive request messages along with the pattern of utilized source ports (if any)
- Fixed values of certain fields within a message that would otherwise be not-fixed
- Refuse to exchange neighbors or botmaster update/attack payloads

Since manual checking cannot always yield a binary answer, i.e., yes or no, the IPs are classified on a best-effort basis using the following classifications: (1) *Highly Possible*, (2) *Possible*, (3) *Unknown*, and (4) *False Positive*. A node is classified as *Highly Possible* when there is significant evidence that resembles a crawler's behavior, e.g., avoiding to exchange information. A node is classified as *Possible* when there is evidence that (almost) equally resembles as both a possible crawler and a bot. A node is classified as *Unknown* when the available evidence is not helpful to make any conclusion. Finally, a node is classified as *False Positive* when logs only indicate the behaviors of a bot. An explanation of why those bots were flagged is also provided.

For the evaluation, an analysis to identify the best threshold values for the parameters t and n_{min} in the frequency-based detection mechanisms, i.e., *SAB-BurstTrap* and *ZAB-BurstTrap*, is first done for both botnets. These threshold values are important to minimize the false positives that may occur due to bots behind NAT and proxy-like devices. Then, the performance BT is evaluated based on the research questions presented in Sect. 4.3.3.3. Finally, the common characteristics exhibited by the detected crawlers are discussed.

4.3.3.3 Research Questions and Expectations

BT detection mechanisms assume an IP address could only be associated to a single crawler or bot. However, the threshold value to trigger a detection in the observed sliding window can be configured in the detection mechanisms (when applicable) to take into consideration influences of bots behind NAT and proxies. Hence, the following research question needs to be answered in the evaluation:

- *What are the suitable threshold values to minimize false positives generated by bots behind NAT and proxies for frequency-based detection mechanisms, i.e., traps within the class of Abuse?*

A BT node can be deployed in most of existing botnets. However, the question remains on how susceptible are current generation of crawlers against such crawler detection mechanisms. Therefore, the following research question needs to be answered in the evaluation:

- *How susceptible are current crawlers against the BT detection mechanisms?*

Considering that not much work has been done in this aspect and the fact that current and previous botnets have only implemented simple crawler detection/prevention mechanisms, it is expected that most of current crawlers are not anticipating such countermeasures. As such, many of the crawlers are expected to be detected by the BT detection mechanisms. However, take note that there might be some crawlers that were left undetected by BT.

Implementation of a crawler can range anywhere from bare minimum to full functionality support of a botnet's protocol (cf. Sect. 2.3.2). Moreover, a crawler can also adopt various strategies to improve its efficiency in crawling the botnets, e.g., multi-threading or distributed crawling. However, very little is known about the characteristics or design choice of the crawlers currently out in the wild. Therefore, the following research question needs to be answered in the evaluation:

- *What are the common characteristics of existing crawlers in the wild?*

4.3.3.4   Results

This section is outlined as following. Firstly, the results of a parameter study for obtaining the threshold values for the BT mechanisms is presented. Then, the evaluation results of the crawler detection mechanisms adapted for Sality and ZeroAccess are presented. Finally, the common characteristics exhibited by the detected crawlers are presented.

Parameter study of suitable threshold values for t and n_{min}

A parameter study of the various combination of parameters of t and n_{min} is conducted. For Sality, the sliding window interval was varied, i.e., 60, 120, ..., 2400 s, and the experiments were repeated with different number of minimum messages required to trigger a detection, i.e., 10, 20, ..., 100 requests. Results indicated that Sality's BT performs best with the parameters $t = 120$ and $n_{min} = 30$. The analysis was also repeated in a similar manner for ZeroAccess, and the results indicated that the best parameters are $t = 60$ and $n_{min} = 40$. The higher number of messages required to trigger detection for ZeroAccess in comparison to Sality is speculated to be contributed by short the MM interval.

Performance of the BoobyTrap mechanisms

The overall results of the detection mechanisms are presented in Table 4.3 according to the respective classes of misbehaviors (cf. Sect. 4.2.2).

For the class of *Defiance*, two BTs were set up for Sality (*SD1* and *SD2*) and one for ZeroAccess (*ZD*). The *SD2-BaitTrap* for Sality was least triggered by crawlers. However, this particular trap is also the most obvious indicator for a crawler as bots in

Table 4.3 Performance of our *BoobyTrap* mechanism. *Source*: https://ieeexplore.ieee.org/stamp/stamp.jsp?tp=&arnumber=7510885

	Defiance			Abuse		Avoidance
	SD1	SD2	ZD	SAB	ZAB	ZAV
Detected IPs	4,212	3	88	11	188	108
After sanitization	966	–	–	–	–	–
Highly possible	4	3	7	9	116	35
Possible	–	–	81	1	72	73
Unknown	962	–	–	–	–	–
False positives	3,246	–	–	1	–	–

Sality would simply ignore entries that are already known and responsive, i.e., a bot would ignore the entry of the BT's secondary port as long as the entry with primary port is still responsive. The *SD1-IgnoreTrap* was triggered by 4, 212 IPs throughout the week, which seems abnormally high compared to other traps. Detailed analysis of the results indicates that many of the flagged IPs are behind ISPs that use multiple NAT IPs or load balancing configurations. Since each request in Sality is sent from a new port (cf. Sect. 3.2.2), NAT devices assume that a new flow or connection is being established and may decide to route the packet using a different proxy or NAT IP as a load balancing technique. As such, the BT node recorded *Hello* messages from a different IP than the one received for the NL_{Req}, thus triggering the trap. These cases were identified and sanitized by correlating a Sality-specific identifier. As a result, we 77% of the detected IPs were identified to be false positives. Out of the remaining 966 IPs, only four IPs exhibited strong indication as crawlers. The remaining 962 IPs could not be reliably classified as their identifiers were set to Sality's default identifier, i.e., 1 (cf. Sect. 3.2.2.1). Hence, they were classified as *Unknown*.

The *ZD-NonComplianceTrap* was triggered by 88 IPs. Seven of those IPs, which were detected on the first day, consistently responded with a *retL* message containing a fixed flag, i.e., $flag = 0$. These IPs are particularly interesting because they responded with exactly 65 messages before they stopped to contact the BT node. It is suspected that these are crawlers that implement a blacklisting mechanism to avoid crawling or contacting other sensors. These IPs were also observed to keep communicating with another instance of sensor node after the BT sensor was (presumably) blacklisted. The remaining IPs responded with a flag value set to either 0 or 1 up to a maximum of five replies. As possibility for such behavior is not observed in the reverse-engineered malware variants (cf. Sect. 3.3), there is no other explanation other than them being crawlers.

The evaluation on the BTs within the class of *Abuse* was conducted based on the frequency of received *NL* request messages for both botnets. The parameters of the BTs were set according to the results of the previous parameter study: Sality ($t = 120$, $n_{min} = 30$) and ZeroAccess ($t = 60$, $n_{min} = 40$). Results indicated 11 flagged IPs by the *SAB-BurstTrap*. Out of the 11, nine IPs were classified as *Highly*

Fig. 4.6 Daily analysis of
SAB-BurstTrap. Source:
https://ieeexplore.ieee.org/
stamp/stamp.jsp?tp=&
arnumber=7510885

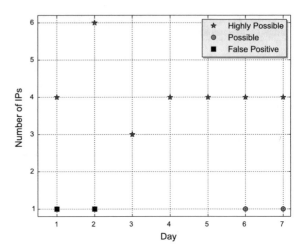

Possible. A daily analysis of this particular BT as presented in Fig. 4.6 indicated
that an average of four crawlers is successfully identified every day. Meanwhile, the
single false positive was identified to be caused by many bots behind a particular
shared IP coincidentally contacting our BT node around the same time.

The *ZAB-BurstTrap* flagged a total of 188 IP addresses throughout the measurement period. After manual inspection, 116 IPs were classified as strongly exhibiting
crawler-like behaviors. The remaining IPs that were classified as *Possible* exhibited
similar behaviors to bots that lack responsive neighbors in their *NL* during their initial bootstrapping phase. This speculation could especially be true considering that a
large portion of the ZeroAccess botnet was sinkholed in 2013 [12]. As such, bots that
experience lack of neighbors could also have requested *NL*s with a higher frequency.

The *ZAV-IgnoreTrap*, as an avoidance trap, within the class of *Avoidance* attempts
to identify crawlers that are refusing to respond to the *crafted* requests sent in the *ZD-
NonComplianceTrap*. The evaluation of this BT indicates 108 IPs in the ZeroAccess
dataset that never responded to any of the request messages. More precisely, 35 IPs
were classified as *Highly Possible* crawlers because the BT node recorded abnormally
high number of received *getL* requests, i.e., between 10 and 14, 800, but without
any replies for any of the crafted request messages. Seventy-three (73) IPs were
classified as *Possible* crawlers. These IPs seemed to be shared among many bots,
i.e., identified by distinct botnet-specific identifiers, but originate only from selected
network prefixes. An alternative hypothesis for this observation can be explained if
there is any packet-level filtering mechanism deployed within those networks that
drops all inbound ZeroAccess' requests or replies. Such a scenario could result in
the BT node observing the behavior of bots *refusing* to respond.

Characterization of detected crawlers

The various detected crawlers were analyzed to identify common characteristics
exhibited by them. The characteristics can be categorized as the following:

- **Blacklisting**: Some crawlers were noticed to be using some blacklisting mechanisms to improve the quality of their crawl data, i.e., ignore sensor nodes. Such crawlers seemed to identify sensor nodes based on the responses received from bots/sensors (explained next), e.g., empty or duplicated NL replies.
- **Sanity Checking**: Some crawlers also seemed to perform sanity-checking on the results obtained from bots. The detection mechanisms within the class of *Defiance* detected a smaller fraction of crawlers than those within the class *Abuse*. For instance, the SD2 detected only three crawlers whereas SAB detected nine. Hence, it appears that some crawlers does follow the implementation of the botnet protocols very closely.
- **Aggressive and/or Persistent Crawling**: Within the frequency-based detection mechanism, only few were observed to crawl continuously, i.e., 24x7. Nevertheless, some of theme crawled the BTs aggressively, i.e., an average 15 requests per minute in the case of Sality.
- **Crawling Redundancies**: Some crawlers were observed to utilize identical botnet-specific identifiers and port numbers across different instances at the same time. This observation may indicate identical crawler instances deployed for redundancies or for obtaining additional vantage points. Multiple vantage points for crawling also help in obtaining a more accurate crawl data if some crawlers suffered network failure or other network-specific issues when crawling.
- **Efficient Crawler Design**: Unlike regular bots, some crawlers were observed to use dedicated a port, i.e., fixed source port, to process incoming NL replies (cf. Sect. 3.3.2). Such a communication design can improve the crawling efficiency as it allows the response-processing thread to be detached from the request-sender thread, allowing the crawler to efficiently crawl more bots in parallel.
- **Identity Hiding**: By checking *WHOIS* information on the detected IP addresses, it is observed that only some of the IP addresses disclose information about the organization or individual that is behind the crawling activities. In fact, some crawlers are observed sharing residential IP addresses, i.e., behind ISP NAT devices. In addition, IPs of some of these crawlers were also observed to constantly change due to dynamic IP address reallocation by ISPs, i.e., IP address aliasing. Such scenarios makes it more difficult to detect crawlers via any frequency-based detection mechanisms.
- **Neutral**: Finally, based on the analysis of the detected crawlers, it can be concluded that the crawlers are neutral. They do not seem to aid the botnets in any manner, e.g., dissemination of botnet commands. Even in cases where neighbors are being returned, these were either other sensors or invalid entries.

4.4 Summary

This chapter presented works on advanced monitoring on the basis of crawling P2P botnets and outlined three major contributions. The first contribution presented a novel crawling algorithm called ZEUSMILKER (see Sect. 4.1.1) that deterministically spoofs keys to request NL from bots to circumvent the GameOver Zeus' NL restriction mechanism. ZEUSMILKER is the first and the only known solution to provably retrieve all NL entries of a GameOver Zeus bot. ZEUSMILKER can circumvent the NL restriction mechanism of GameOver Zeus by exploiting the deterministic neighbor selection mechanism and the fact that keys included in the NL requests can be spoofed. Concluding, both above-mentioned factors have allowed this anti-crawling mechanism to be circumvented by ZEUSMILKER.

To anticipate the retaliation of the botmasters against ZEUSMILKER, Sect. 4.2.1 proposed several enhancements, i.e., *Random Node Return*, *Bit-XOR+*, and *Bit-AND*, to the existing GameOver Zeus NL restriction mechanism. These proposed countermeasures address the drawbacks of the original mechanism: preventing the requester to manipulate the returned entries. The evaluation of the new proposals indicated that *Bit-AND* performs best compared to the other two proposed mechanisms in impeding the performance of crawlers. However, this countermeasure adversely affects the resulting botnet overlay, so it is not likely to be used. Therefore, the *Bit-XOR+* is most likely to be adopted by future botnets. Both above-mentioned mechanisms can affect the ability of a crawler to retrieve the (complete) NL of a bot, and therefore need urgent attention of the researchers as future botnets may adopt these mechanisms.

The second contribution in this chapter proposes a novel crawling algorithm called *LICA* (cf. Sect. 4.1.2) that attempts to enumerate as many bots as possible economically. LICA attempts to approximate a *minimum vertex cover* that represents the minimum set of bots that need to be crawled to discover all bots in the botnet. By prioritizing *popular* bots that are returned by other bots, this algorithm crawls the backbone of the botnet and can terminate as soon as the ratio of newly discovered bots falls under a certain threshold value. Evaluation results indicated that LICA outperforms other state of the art crawling algorithms, in particular BFS and DFS-based graph traversal techniques. This algorithm can be utilized in crawling P2P botnets in a more stealthy manner.

As the third contribution, a lightweight crawler detection mechanism called BoobyTrap (BT) that exploits botnet-specific protocol and design constraints was proposed. BT aims at detecting crawlers in an autonomous manner by analyzing the communication of other bots with itself. Based on simple test-cases, a behavior of a crawler can be distinguished from bots. Evaluation results in Sect. 4.3.3 indicated that many crawlers in Sality and ZeroAccess can already be detected by BT.

The findings of the different work presented within this chapter imply that more advanced monitoring mechanisms are needed to tackle future P2P botnets. Additionally, such advanced mechanisms should focus on the following:

- **Larger pool of IP addresses**: Since most of the existing and proposed anti-crawling mechanisms are based on the fact that an IP address represents a crawler, it is important to acquire a larger pool of IP addresses that can be used for future crawling activities. Most importantly, these IP addresses should not be of a single contagious block or range of IP addresses to avoid bots applying IP prefix-based blacklisting similar to that implemented by GameOver Zeus. This way, even if some of the IP addresses were blacklisted, other IP addresses can serve as redundancies to continue monitoring. Such non-contagious block of IP addresses could also be obtained through the cooperation of several interested organizations or institutions.

- **Distributed crawling :** Future botnet monitoring activities should also consider using distributed crawlers in combination with a large pool of IP addresses to circumvent IP-based anti-crawling mechanisms, e.g., BT or NL restriction mechanisms, to capture the characteristics of bots accurately. Hence, crawlers can easily circumvent any IP address or frequency-based anti-crawling mechanisms.

References

1. Rossow, C., Andriesse, D., Werner, T., Stone-gross, B., Plohmann, D., Dietrich, C.J., Bos, H., Secureworks, D.: P2PWNED: modeling and evaluating the resilience of Peer-to-Peer botnets. In: IEEE Symposium on Security and Privacy (2013)
2. Karuppayah, S., Roos, S., Rossow, C., Mühlhäuser, M., Fischer, M.: ZeusMilker: circumventing the P2P zeus neighbor list restriction mechanism. In: IEEE International Conference on Distributed Computing Systems (ICDCS) (2015)
3. Karuppayah, S., Fischer, M., Rossow, C., Mühlhäuser, M.: On advanced monitoring in resilient and unstructured P2P botnets. In: IEEE International Conference on Communications (ICC) (2014)
4. Maymounkov, P., Mazieres, D.: Kademlia: a peer-to-peer information system based on the xor metric. Peer-to-Peer systems. Lect. Notes Comput. Sci. **2429**, 53–65 (2002)
5. Stutzbach, D., Rejaie, R., Sen, S.: Characterizing unstructured overlay topologies in modern P2P file-sharing systems. ACM SIGCOMM Internet Meas. Conf. (IMC) (2005)
6. Bar-Yehuda, R., Even, S.: A local-ratio theorem for approximating the weighted vertex cover problem. Ann. Discret. Math. (1985)
7. Falliere, N.: Sality: Story of a Peer-to-Peer Viral Network. Technical report, Symantec (2011)
8. Karuppayah, S., Vasilomanolakis, E., Haas, S., Mühlhäuser, M., Fischer, M.: BoobyTrap: on autonomously detecting and characterizing crawlers in P2P Botnets. In: IEEE International Conference on Communications (ICC) (2016)
9. Rossow, C.: Amplification hell: revisiting network protocols for DDoS abuse. In: Network and Distributed System Security Symposium (2014)
10. Baumgart, I., Heep, B., Krause, S.: Oversim: a flexible overlay network simulation framework. In: IEEE Global Internet Symposium (2007)
11. Hagberg, Aa, Schult, Da, Swart, P.J.: Exploring network structure, dynamics, and function using NetworkX. In: Proceedings of the 7th Python in Science Conference (SciPy2008), **836**, 11–15 (2008)
12. Neville, A., Gibb, R.: ZeroAccess Indepth. Symantec Security Response (2013)

Chapter 5
Deployment of Sensor Nodes in Botnets

The deployment of sensors within P2P botnets allows additional vantage points in monitoring bots. In particular, sensors can enumerate bots that are otherwise not discoverable by crawlers due to their inability to contact bots behind network devices like NAT and stateful firewalls (see Sect. 2.3.2).

A sensor node is deployed in a botnet by *announcing* its presence to other super-peers leveraging the node announcement mechanism of the botnet (see Sect. 1.1.3). Sensor announcements are often carried out by crawlers that (in)directly announce the presence of the sensor during crawling. After being well-known amongst many super-peers, the information of the sensor will be frequently handed out to non-superpeers that may request additional neighbors from existing superpeers. Thereafter, a sensor node receives increased communication requests from non-superpeers that have added the sensor as a candidate in their NL. As a consequence, an effective sensor is usually very *popular* amongst all bots in a botnet, i.e., known by many bots.

By combining the monitoring data of both crawlers and sensors, a more accurate enumeration of a botnet's population can be obtained for further analysis. In addition, a variation of a sensor is often used as a *sinkhole* server in botnet takedown attempts. Such sinkhole servers would usually aim to remain responsive to probing messages of bots as outlined in Sect. 2.2. In addition, the sinkhole servers would not disseminate any new botnet updates or commands to prevent the bots from communicating with the botmaster. Therefore, sensors do not only pose as a threat to botnets due to its monitoring capabilities but also as a tool or stepping stone to launch botnet takedown attacks. Nevertheless, not much work has been done in the area of preventing or detecting sensors in P2P botnets.

Unlike the content organization of Chap. 4, due to the lack of prior work, this chapter starts from a perspective of a botmaster in Sect. 5.1 to introduce three mech-

The original version of this chapter was revised: Belated corrections have been incorporated. The erratum to this chapter is available at https://doi.org/10.1007/978-981-10-9050-9_7

S. Karuppayah, *Advanced Monitoring in P2P Botnets*, SpringerBriefs on Cyber Security Systems and Networks, https://doi.org/10.1007/978-981-10-9050-9_5

anisms to detect sensor nodes deployed in botnets. Then, from the perspective of a defender, Sect. 5.2 proposes countermeasures to circumvent the detection mechanisms. Finally, Sect. 5.4 concludes this chapter.

5.1   Detecting Sensor Nodes in Botnets

As indicated by the lack of prior work, there are many challenges in detecting sensor nodes deployed in P2P botnets. In contrast, this section will show that it is indeed possible to detect sensor nodes by relying upon graph-theoretic metrics of the botnet overlay. First, some introduction is presented to understand on why is it challenging to detect sensors. After that, the detection mechanisms are detailed.

5.1.1   Introduction

This subsection provides some basic introduction on sensors and the issues in detecting them. Particularly, Sect. 5.1.1.1 introduces and discusses the most important feature of a sensor node: handling of botnet request messages. Section 5.1.1.2 briefly describes the process of deploying the sensor node in a P2P botnet. Section 5.1.1.3 presents a set of assumptions for a sensor node detection mechanism that were derived from the discussions of Sects. 5.1.1.1 and 5.1.1.2 and own observations on sensors deployed in existing botnets. These assumptions are important and are the basis for the remaining work presented in this chapter. Finally, Sect. 5.1.1.4 discusses the challenges often faced in detecting sensor nodes.

5.1.1.1   Message Handling by a Sensor Node

The main functionality of a sensor node is the handling of botnet-specific communication messages, i.e., responding request messages with valid replies. A sensor is usually only required to handle a few relevant messages for the purpose of botnet monitoring. Some of the most common types of messages are detailed in the following:

1. **(Responsiveness) Probe Messages**: *probeMsg* is usually sent by bots to assert the responsiveness of a particular bot (see Sect. 2.2). If a sensor node fails to respond to these messages, bots will eventually flush the entry of the sensor from their NL. Therefore, a sensor needs to always be able to handle and respond to the messages.Moreover, by remaining in the NL of many bots, the information about the sensor can be quickly propagated to newer bots that may request potential neighbors.
2. **NL Request Messages**: *requestL* is usually sent by bots that require additional neighbors (see Sect. 2.2). Although this message handling is often optional for a sensor node, it does help the sensor not to raise suspicions if the handling is

not implemented. A sensor that deliberately ignores handling or replying such messages may alert an attentive botmaster. Based on observations, there are some sensors that either do not respond at all to such request messages, i.e., ignoring the messages, or respond with just an empty but valid reply. The second approach of returning empty replies can be more suspicious although such replies are also valid in some botnet protocols, e.g., GameOver Zeus and Sality.

3. **Botmaster Command Request Messages**: This type of message is sent by bots to query if there are any newer updates from the botmaster that can be downloaded by the bots, e.g., *Hello* message for Sality or *VersionRequest* for GameOver Zeus. However, all of the three analyzed botnets integrate the functionality of this message along with the *probeMsg* described earlier. Bots use sequence numbers to indicate the most current botmaster command known to them. Whenever a bot shares the information that it knows of a newer command, other bots will try to pull the latest update. As such, the probe for the responsiveness of a bot also checks if its neighbor has any new updates.

5.1.1.2 Popularizing the Sensor Node

Deploying a sensor within a botnet is often done by leveraging the node announcement mechanism of the botnet to *popularize* the sensor. Thereafter, popularization of a sensor is usually performed in tandem using a crawler. Aggressive popularization strategies such as *Popularity Boosting* by Yan et al. [9] can be detected by crawler detection mechanisms such as the BT mechanism (see Sect. 4.2.2). However, a non-aggressive popularization strategy is sufficient to deploy the sensor in most botnets. The only drawback with a slow popularization technique is the fact it takes longer before most bots discovers the sensor.

5.1.1.3 Assumptions for a Sensor Node Detection Mechanism

In the following, a set of assumptions for a sensor detection mechanism is presented based on the discussions in Sects. 5.1.1.1 and 5.1.1.2. These assumptions were derived from own observations of the sensor deployed in the wild as well as to adhere to legal requirements.

1. **A sensor is already deployed in the botnet**. Since sensors can be deployed and popularized using less aggressive strategies, detection mechanisms need to focus on detecting sensors that are *already* deployed; those that have evaded BT-like detection mechanisms.

2. **The sensor node does not return any valid bots as neighbors**. A sensor should adhere to legal requirements by not aid the bots in the regular maintenance of the botnet overlay or malicious activities of the botnet. Hence, a sensor should not return any valid bot information when being requested. Instead, the node can either not respond to the message, or return information of other sensors, or invalid entries.

3. **The sensor does not disseminate or exchange any valid update or command from the botmaster with other bots**. This assumption is relevant in the context of adhering to the legal requirements by not participating in a botnet related maintenance. Instead, a sensor node should avoid returning any valid command or update that may benefit the botnet. This can be easily achieved by bluffing on the latest command that is known to the sensor or by not responding.

4. **The total number of sensors under the control of any attacker is lesser than the total number of bots**. An attacker, e.g., a researcher, is assumed to deploy only lesser number of (colluding) sensor nodes than the total number of bots in the botnet, i.e., <50%. This assumption holds because it usually does not require too many sensor nodes to be deployed to monitor a botnet. Furthermore, any party that might have deployed more sensors than the total bots would have already tainted most of the botnet monitoring data rendering any collected monitoring data useless.

5.1.1.4 Challenges in Detecting Sensors

The main challenge in detecting a sensor is to distinguish it from regular bots [1]. Sensor nodes are usually very popular among bots, i.e., having high indegree. However, this behavior is similar with superpeers that have been reliable for a long period. Moreover, the passive nature of a sensor that generates only minimal network traffic as it only responds to incoming requests, in comparison to crawlers that actively generate requests at high frequencies [1], makes a sensor more difficult to be distinguished. In contrast, the work presented in Sects. 5.1.2–5.1.4 will show that connectivity metrics can be used to distinguish sensors from bots.

5.1.2 Local Clustering Coefficient (LCC)

This detection mechanism attempts to detect sensors based on the inter-connectivity relationship amongst neighbors of a node. This detection mechanism exploits the observation that in unstructured P2P botnets, nodes with high uptime tend to establish neighborhood relationships among themselves and thus form a backbone. As backbone nodes are popularly contained in the NLs of most bots in the botnet, there is a high probability that an ordinary bot has several backbone bots in its NL.

The degree of inter-connectivity of the neighbors of a bot can be represented by the *clustering coefficient* metric. The clustering coefficient is often used to express the density of networks. In this work, the *Local Clustering Coefficient (LCC)* introduced by Watts and Strogatz [8] is used to express the connectivity of a node's neighbors by computing their degree of inter-connectivity. Extreme values of 0.0 and 1.0 indicate that the neighbors are either *not* connected at all amongst each other or that they are completely *mesh*.

To detect sensors, a snapshot of the botnet overlay is required. The *directed* variant of the LCC is calculated using these snapshots for each bot x, $lcc^+(x)$, to analyze the inter-connectivity of its neighbors by using Eq. (5.1). E is the set of all edges in the network and NL_x represents the NL of a bot x. The mechanism sets $lcc^+(x) = 0.0$ if $|NL_x| = 0$ or 1 (the numerator will fast-evaluate to 0)

$$lcc^+(x) = \frac{|\{(u, v) \in E : u, v \in NL_x, u \neq v\}|}{|NL_x| \times (|NL_x| - 1)} \tag{5.1}$$

Figure 5.1 presents the analysis of LCC on Sality V3 for a given snapshot on the LCC values of nodes in dependence to their popularity, i.e., indegree. It is clear that most bots exhibit a similar degree of inter-connectivity in their neighborhood due to the presence of common set of backbone nodes in their NLs, i.e., a majority of the bots having $0.6 <= lcc^+(x) < 0.8$. However, according to the assumptions laid out above, sensors do not share valid bots when they receive a NL request. Thus, their lcc^+ would differ from bots.

As depicted in Fig. 5.2, a sensor has three possible behaviors upon receiving a NL-request as explained in Sect. 5.1:

1. Return *no* neighbors or ignore the request. This behavior will lead to $lcc^+(x) = 0.0$, e.g., S_A in Fig. 5.2.
2. Return only *invalid* neighbors. As invalid neighbors are not inter-connected or reachable, again $lcc^+(x) = 0.0$ holds, e.g., S_A in Fig. 5.2.
3. Return only *responsive sensors*. If each of the returned sensors return each other as their neighbors, $lcc^+(x) = 1.0$ will hold since the sensors are in a full mesh, e.g., S_B, S_C, or S_D in Fig. 5.2. However, if the connectivity among the returned sensors is as in a directed cycle or not connected at all, it will lead to $lcc^+(x) = 0.0$.

As depicted in Fig. 5.1, due to the tendency of having common set of backbone nodes in the NLs, regular bots will not have these extreme lcc^+ values. Therefore, nodes exhibiting extreme values can be flagged as potential sensors by LCC.

5.1.3 SensorRanker

The second sensor detection mechanism is called *SensorRanker*. It uses the *PageRank* algorithm [4] to distinguish sensors from bots. The PageRank algorithm was initially

Fig. 5.1 LCC values of bots in Sality V3. *Source*: https:// ieeexplore.ieee.org/stamp/ stamp.jsp?tp=& arnumber=7346908

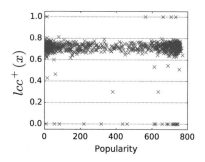

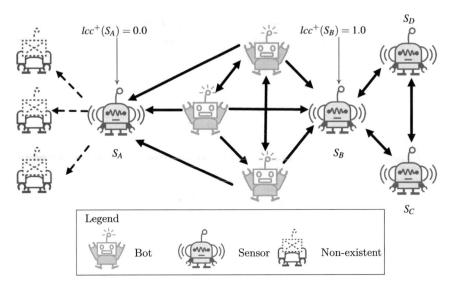

Fig. 5.2 Extreme values of LCC can be used to identify sensors deployed within a botnet, i.e., $lcc^+(x) = 0.0$ or 1.0

designed to determine the importance, i.e., popularity, of websites based on the number of pages referring to them via hyperlinks. Towards this, a relation of websites and hyperlinks is modeled as directed graphs with the websites as *nodes* or *vertices* and the hyperlinks as *edges*. This botnet formal model is extended in the following to describe PageRank and SensorRanker in more detail.

Extended Botnet Formal Model

The neighborhood relationship, i.e., *NL*, of a peer $v \in V$ can be defined as the successors of v, $succ_v = NL_v = \{u | \forall u \in V : (v, u) \in E, v \neq u\}$ that contains the set of all peers to which v has an outgoing connection. The NL can also be more specifically expressed as NL_v^t to reflect the exact view, i.e., outdegree, of the NL of peer v at time t. Consequently, the set of bots that have bot v as their neighbors or incoming connections to v can be expressed by the set of predecessors $pred_v = \{u | \forall u \in V : (u, v) \in E, v \neq u\}$, i.e., indegree.

The PageRank algorithm assigns values between 0.0 and 1.0, where a higher value denotes higher *rank* or popularity of a node v, e.g., $PR_v = 1.0$. The values are calculated based on a node's predecessors $pred_v$ and their respective ranks. In each iteration of the algorithm, the rank of a node is distributed equally among all its outgoing edges, i.e., $succ_v$. The rank-value distributed over all edges of a node v is expressed as $edgeweight_v = \frac{PR_v}{|succ_v|}$. The PageRank value of a node, in turn, is the *sum* of the edge-weights of all of its predecessors.

The concept of ranks in PageRank is also directly comparable to the popularity of bots in a P2P botnet. Bots become more widely known and popular in the botnet when they have been available and responsive for a prolonged period. However, sensors are also equally popular when they are widely known amongst many bots. As such, popularity alone is not sufficient nor effective to distinguish sensor from popular bots [1]. However, when taking PageRank into consideration, the edge-weights on outgoing edges for sensors differ greatly than popular bots because they have, either none or very few outgoing edges compared to the popular bots, i.e., sensor nodes have higher edge-weights due to very few (if any) out going connections. This discrepancy can be exploited to distinguish sensors from bots using the edge-weight as a reliable metric.

Nevertheless, due to the churn dynamics in P2P botnets, using the original PageRank algorithm as it is may indicate unpopular bots, i.e., bots known only by a small fraction of superpeers, that have some predecessors that are coincidentally with very high PageRank values, to appear as having a very high edge-weight. The edge-weight of a node v is normalized to address this drawback by multiplying it by its *popularity ratio*, i.e., ratio of predecessors over the size of the botnet population. This adapted PageRank algorithm is the proposed *SensorRank* value and is defined as:

$$SensorRank_v = edgeweight_v \times \frac{|pred_v|}{|V|} \tag{5.2}$$

Although the SensorRank values for sensors would be significantly higher than those of bots, a means to automate the detection of sensors is still needed from the perspective of a botmaster. For this, clustering algorithms from the domain of machine learning are utilized to assist in distinguishing sensors from bots. The details of the clustering algorithms are elaborated later in Sect. 5.3.2.

5.1.4 SensorBuster

The third proposed mechanism is called *SensorBuster*. It utilizes the *Strongly Connected Component (SCC)* connectivity metric introduced by Robert Tarjan [7] to identify sensors. An SCC of a directed graph G is defined as a maximum set of vertices $C \subseteq V$ with a directed path between each pair of nodes $(u, v) \in C$, i.e., $u \to v$ and $v \to u$.

Considering that P2P botnets rely heavily on the inter-connectivity among bots to prevent segmentation or partitioning of the overlay, bots $B \subseteq V$ often form a *single* SCC where there is a path to and from one bot to another. From here onward, such an SCC is referred to as the *main* SCC. Please note that from the fourth assumption presented in Sect. 5.1.1.3, the largest SCC can be safely assumed as the main SCC, i.e., the SCC formed by bots are larger than those formed by sensor nodes. Without the main SCC, any new command disseminated in a botnet would require a longer period to reach all bots. Due to the assumptions about sensors presented in Sect. 5.1.1.3, a sensor would not be part of this main SCC, since a sensor node will not have any bot as its successor, i.e., no paths from the sensor into the main SCC. Therefore, sensors will form their own SCC consisting of either only a single sensor or multiple colluding sensors. As such, all nodes that are not included in the main SCC are most likely sensors (see Fig. 5.3).

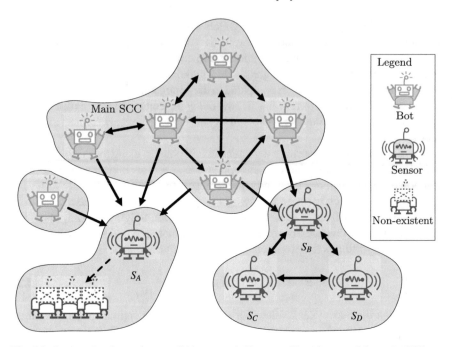

Fig. 5.3 Sensors that do not return valid bots as neighbors would not be part of the *main* SCC

5.2   Circumventing Sensor Detection Mechanisms

The previous section proposed three detection mechanisms to detect sensors. In this section, from the perspective of a defender, methods to circumvent the sensor detection mechanisms as discussed in Sect. 5.1 are proposed. All proposed methods in this section require a set of colluding sensors to circumvent the detection mechanisms. Specifically, sensors utilize a distributed sensor deployment strategy called *Distributed Sensor Injection (DSI)*. In this strategy, sensors controlled by a user are used to manipulate the observed connectivity metrics by intelligently distributing loads among multiple colluding sensors.

The DSI strategy assumes the following:

1. At least four colluding sensors are available, i.e., $|S| \geq 4$.
2. Colluding sensors communicate with the user using out-of-band communication channels.
3. The user can instruct the sensors to ignore communications from selected bots or attempt to *inject* or *announce* themselves into the NL of other bots.

The remainder of this section is outlined as follows: Sect. 5.2.1 introduces a method to manipulate LCC. Meanwhile, Sect. 5.2.2 presents a method to circumvent SensorRanker. Finally, Sect. 5.2.3 discusses methods to evade SensorBuster.

5.2.1 Circumventing LCC

LCC calculates the clustering coefficient of each bot v and identifies sensors that have extreme values of $lcc^+(v) = 0.0$ or $lcc^+(v) = 1.0$. Such values indicate that neighbors of a bot v are not connected at all or connected in a full mesh. As the majority of bots have common neighbors in the form of reliable backbone nodes [2], it is unlikely that a bot could have neighbors that are not inter-connected at all, i.e., $lcc^+(v) = 0.0$. Therefore, it could only be of sensors that refuse to share any neighbors or those that shared non-existent neighbors.

Moreover, due to large NL sizes in botnets such as Sality and ZeroAccess, i.e., $|NL| \geq 256$, it is also unlikely that all neighbors seen in NL have exactly each other in their NL. Nodes having each other in their NL is more likely the case of a group of sensors attempting to popularize themselves.

Since LCC detects only extreme values, it can be easily circumvented by having sensors not exhibiting these extreme values. By using a minimum of four sensors that are connected among themselves as depicted in Fig. 5.4, each sensor can intelligently yield a clustering coefficient of $lcc^+(S_i) = 0.5$. The main idea is to avoid having a full mesh connectivity between the sensors but at the same time having some common connectivity with other sensors, i.e., $lcc^+(v) \neq 0.0$ and $lcc^+(v) \neq 1.0$. Therefore, for a group of N sensors where $N > 3$, a user needs to connect each sensor S_i, $i \in [0, N]$ to all other sensors except $S_{i-1 \mod N}$ and to the sensor itself, i.e., to avoid self-loop.

5.2.2 Evading SensorRanker

The SensorRanker mechanism requires a more sophisticated mechanism to evade detection. SensorRanker focuses on the popularity of nodes; that is defined by their indegree. As such, the more popular a bot is, the higher the SensorRank value is. Therefore, it is important that sensors avoid being *abnormally* popular. Moreover, having sufficient outgoing edges help to reduce the SensorRank value of a node.

Fig. 5.4 A minimalistic example of colluding sensors evading the LCC mechanism

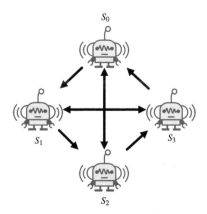

By using the DSI strategy, a user can distribute the popularity of his sensors among the set of sensors in an optimal manner, i.e., distributed evenly. For instance, to obtain the full enumeration of the botnet's population $|V|$, one could distribute N sensors to have $\frac{|V|}{N}$ predecessors. It is also important that the set of predecessors of a sensor S_i is to be distributed evenly among all sensors based on the *edge-weight* of the bots that would also influence the SensorRanker detection mechanism (cf. Sect. 5.1.3). Hence, sensors not only reduce their popularity significantly but the increased number of interconnections with other sensors also further reduces the *SensorRank* value of a sensor.

$$SR(S_i) = \frac{PR(S_i)}{N_{min} - 2} \times \frac{1}{N_{min}} \leq Avg(SR(V - S)) \qquad (5.3)$$

To circumvent the SensorRanker mechanism, one additional assumption is adopted: SensorRanker can be evaded if the SensorRank value of a sensor S_i is lower than or equal to the average value of all bots, i.e., $SR(S_i) \leq Avg(SR(V - S))$. As such, the *minimum* amount of sensors needed to circumvent SensorRank, i.e., $N_{min} = |S|$, can be calculated by satisfying Eq. (5.3). This equation calculates the *SensorRank* value of a sensor S_i depending on the number of outgoing edges, i.e., other sensors, to satisfy the condition of being lesser or equal to the average value of all bots. However, depending on a botnet's *NL*-size, this method could be resource-intensive or costly regarding the numbers of sensors needed to evade this mechanism. Detailed analysis through a simulation study on the feasibility of circumventing the SensorRanker mechanism is presented in Sect. 5.3.4.

5.2.3 Evading SensorBuster

The DSI strategies proposed above will still be detected by SensorBuster as the colluding sensors create their own SCC with no connection back to any other bots. Evading SensorBuster requires, at least, one connection from any of the sensors, to and back from the main SCC. However, to the best of knowledge, there are no possible strategies to evade this mechanism, unless an assumption that is presented in Sect. 5.1.1.3 is ignored, i.e., handing out bots as neighbors when being requested.

Therefore, for the sake of completeness, this assumption is ignored in the context of evading SensorBuster. By ignoring the assumption, the proposed DSI strategies not only can circumvent SensorBuster but also all other detection mechanisms, i.e., LCC and SensorRanker. Therefore, that particular assumption is relieved so that handing out legitimate neighbors would be permitted for sensors.

The following method circumvents the detection mechanisms by returning bots when being requested. However, this method also attempts to ensure that the botnet receives only *minimal* benefits from the returned bot. Stutzbach and Rejaie reported that nodes in P2P networks which already exhibit a long *uptime* would most likely continue to remain available [6]. In contrast, nodes that are newly seen have a higher probability of leaving the network immediately. Since the observation can also be generalized to P2P botnets, only neighbors that have recently joined would be picked and returned when being requested. Such bots have higher tendency to be less useful, i.e., most likely to go offline soon. One could also argue that well-known bots should be returned instead of newer bots. However, well-known bots usually form the backbone of a botnet overlay and are very important to the maintenance of the botnet overlay. In contrast, returning newer ones allows the SensorBuster mechanism to be circumvented with a very minimum assistance *offered* to the botnet.

For this, each sensor would be required to keep track of two timestamps of each discovered superpeer, i.e., *firstSeen* and *lastSeen*.Then, sensors may choose to alternate between returning other sensors and sometimes return legitimate neighbors by picking bots with the most recent *firstSeen*. By returning bots, sensor(s) would no longer form an isolated strongly connected component but merge with that of the whole botnet itself since there is a path to and from the main component and hence circumvents SCC.

5.3 Evaluation

This section presents the evaluation results and analysis of the mechanisms proposed in Sects. 5.1 and 5.2 as outlined in the following. Section 5.3.1 describes the datasets utilized for evaluating the sensor detection mechanisms. Then, Sect. 5.3.2 elaborates the setup for the experiments. Section 5.3.3 discusses the investigated research questions as well as the expected outcomes. Finally, Sect. 5.3.4 presents the results of the experiments.

5.3.1 Datasets

The datasets were obtained by continuously crawling the Sality (Version 3) and ZeroAccess botnets for a duration of one week, respectively. Sality was crawled from 08/10/2015 00:00:00 UTC to 14/10/2015 23:59:59 UTC. Due to the neighborlist return mechanism in Sality that returns only one entry when requested (cf. Sect. 3.2.2), a multi-session crawling that sends 30 simultaneous requests to each bot for every crawl session was conducted. Please note that for Sality, a *Hello* request message is additionally sent to each bot before crawling the bot for every crawl session. A response to the sent message allows the crawler to assert the responsiveness of a bot before crawling it. The ZeroAccess botnet was crawled from 07/11/2015 00:00:00 UTC to 13/10/2015 23:59:59 UTC with a single request message that yields 16 neighbors for every crawl session.

From the initial 32,693 bots discovered in Sality, 4,131 bots have been pruned because they have never responded to any *Hello* messages, i.e., suspected artifacts resulting from bots that went offline. Similarly, out of 95,668 bots discovered in ZeroAccess, 93,632 bots have been removed, because they never responded to

Table 5.1 Summary of the sanitized datasets

	Sality (Version 3)	ZeroAccess
Total bots	28,562	2,306
Hourly Avg. (bots)	1,479	105
Max. Neighbors	656	134
Min. Neighbors	0	0
Avg. Neighbors	318	86
Median Neighbors	369	93

requests, i.e., suspected artifacts resulting from bots that went offline and an ongoing active pollution attack. The summarized details of both sanitized datasets are presented in Table 5.1.

5.3.2 Experimental Setup

The performance evaluation experiments utilized *Python* scripts that were built upon the *NetworkX* [3] and *scikit-learn* [5] modules to implement LCC, SensorRanker, and SensorBuster. The detection mechanisms require the crawl data as an input to detect sensors on the basis of the connectivity characteristics. However, several potential issues will surface in both using the snapshots and to accurately detecting sensor nodes. The following subsection addresses these concerns:

As it is difficult to capture the 'complete' state of a bot's NL at a given point in time, Sect. 5.3.2.1 suggests a method to split crawl sessions into smaller chunks and use them as a representation of a bot's NL instead. Section 5.3.2.2 discusses how auxiliary data obtained during crawling can be used to help to assert a detection of a sensor or help to identify false positives. Meanwhile, Sect. 5.3.2.3 presents a mechanism to identify and remove artifacts within the obtained snapshots that could otherwise adversely affect the performance of the detection mechanism.

5.3.2.1 Splitting Crawl Data into Snapshots

Since the *NL*-reply mechanism that is adopted by both Sality and ZeroAccess prevents a crawler from capturing the complete NL_v^t of a bot v at time t, an approximation or a near-complete representation is required as a replacement. Therefore, the results from the multiple crawl sessions are aggregated into hourly snapshots to represent as the *near-complete* NL of bots at any given hour $t \in [1, 24]$.

Each snapshot in the Sality dataset has an average of 81 crawl sessions, which corresponds up to about 2,430 requests sent to each bot within an hour. As for ZeroAccess, each snapshot in the dataset has an average of 299.7 crawl sessions. Each of these snapshots is considered as a "complete" botnet topology at the given point of time, i.e., hours. Only one snapshot per day is chosen and selected based

Table 5.2 Summary of the selected snapshots

	Sality (Version 3)	ZeroAccess
Total bots	3,975	325
Hourly Avg. (bots)	1,061	91
Max. Neighbors	564	111
Min. Neighbors	0	0
Avg. Neighbors	340	72
Median Neighbors	434	77

on the one with the lowest number of bots seen in a day. Since sensors always attempt to be responsive (cf. Sect. 5.1), it is inherently assumed that a sensor would be responsive throughout every hour of a given day. Therefore, it is sufficient to execute the detection mechanisms on the snapshot with the least nodes as such a snapshot reduces the probability of increased false positives or artifacts (if applicable). From the analysis of the hourly snapshots, the 5th snapshot of any day, i.e., 04:00:00–04:59:59, is the lowest for Sality. In comparison, it was the 7th snapshot of any day for the ZeroAccess botnet. Details of the seven selected snapshots for each botnet is presented in Table 5.2.

These snapshots which are used as inputs for each of the detection mechanisms are hereafter referred to as the respective botnet's dataset itself. After running the detection mechanisms on the input datasets, each mechanism will generate a list of IPs that are flagged as potential sensors.

5.3.2.2 Using Auxiliary Data to Help in Making Decisions

Although the sensor detection mechanisms can flag potential sensors, it would be important for a botmaster to inspect the flagged nodes further before deciding if they are indeed true positives or false positives. For that, all metadata and payload contents of each received response should be logged by the crawler.

Strengthening Confidence

Bots in Sality that receive a *Hello* message with an older *URLPack* would respond by attaching the latest *URLPack* known to them. Hence, by transmitting an older sequence number of the *URLPack* within the sent *Hello* messages before crawling a bot, the corresponding response should consist of an attached recent *URLPack* from the bots. Similarly, all *getL* messages that are sent to a ZeroAccess bot should also be responded with a *retL* message that would consist of all *plugins* that are available for download from the responding bot. The details of which *URLPack* or *plugin* is missing within the responses are logged as part of the auxiliary data. In addition, the neighbors returned by bots in Sality and ZeroAccess for each received NL_{Rep} or *retL* is also logged.

Based on the auxiliary data, it is possible to strengthen the confidence of accurately flagging a sensor by inspecting if there were any logged misbehavior, e.g., missing

URLPack or *plugins*. Although the usage of the auxiliary data itself can be used as a technique to identify sensors, there are an arbitrary number of reasons that can mislead this detection technique, e.g., network-specific anomalies. For instance, Andriesse et al. reported that they were not able to observe any node that refused to exchange the *URLPack* in Sality [1]. In contrast, a more current analysis conducted within the scope of this work indicated that there were indeed nodes that were refusing to exchange their *URLPack* when requested. Hence, the collected data is only used as a reference to further *strengthening* the confidence of any flagged nodes by the proposed detection mechanisms.

Identifying False Positives

It is also possible to identify false positives due to temporal network issues experienced by bots based on historical data. For instance, all neighbors of a bot could coincidentally be offline at the same time and forced the bot to exhibit the behavior of not having any valid neighbors to share. However, by looking at past or future historical records of the particular bot, it is possible to identify such a scenario and mark the detection as a false positive correspondingly.

Classifying Flagged Nodes

Based on the collected auxiliary data, nodes flagged by the detection mechanisms can be classified into the following categories:

1. **Sensor**: Nodes exhibiting characteristics that correspond to the assumptions in Sect. 5.1
2. **False Positive**: Nodes exhibiting characteristics that conform to the botnet protocol
3. **Unknown**: Nodes that are not able to be classified as either *Sensor* or *Bot* due to lack of information

5.3.2.3 Handling of Churn Artifacts

Since the detection mechanisms heavily depend on connectivity-specific metrics, artifacts introduced during crawling may influence the accuracy of the detection mechanisms. Two different strategies can be adopted to remove such artifacts from the datasets. The first strategy is to remove bots that have no neighbors at all, i.e., bots that have never responded to any NL-request messages. However, this strategy is not recommended as it may also remove sensors that are designed to ignore such messages.

The second strategy is to measure and utilize the *responsiveness* of a bot to identify and remove artifacts. The *responsiveness* of a bot v at time t can be expressed as the ratio of the number of received replies to the number of sent requests between $t - \delta t$ and t:

$$R_v^t = \sum_{\tau=t-\delta}^{t} \text{rep}_v^\tau \times \frac{1}{\sum_{\tau=t-\delta}^{t} \text{req}_v^\tau} \tag{5.4}$$

The responsiveness for bots in Sality is measured based on the number of received *Hello* replies, while for ZeroAccess the number of received *retL* messages. By specifying the *minimum* ratio of responsiveness that is required for any bot within a given snapshot, poorly responsive bots can be identified as artifacts or churn affected nodes. For instance, a value of $R = 0.4$ represents all bots that have been responsive at least 40% of the whole monitoring period, i.e., the total number of crawl sessions within that particular snapshot. Take note that a value of $R = 0.0$ is used to represent bots that have responded at least once to the probe requests. Sensors usually do not get flagged as artifacts as they would aim to be responsive to the probing messages as much as possible unless they are experiencing poor network connectivity. Therefore, all detection mechanism would simply ignore nodes without a minimum responsiveness threshold value from their final classification.

5.3.3 Research Questions and Expectations

In the following, research questions that focus on the evaluation of the sensor detection mechanisms are presented along with investigated parameters.

An important aspect of any detection mechanism is the need to specify the baseline or ground truth. Hence, the following research question needs to be answered in the evaluation:

• *How to establish ground truth on the total number of sensors present in a datasets?*

Since it is difficult to establish such ground truth in botnets, the SensorBuster mechanism is used to provide a baseline information on the maximum number of sensors present in the datasets. The simplistic design of SensorBuster allows it to detect sensor nodes that adhere to the assumptions presented in Sect. 5.1.1.3. Since sensor nodes are assumed to not return a bot as a neighbor, sensors would establish isolated *strongly connected component(s)* which will not have any bots within them. Hence, all IPs flagged by this mechanism needs to be inspected, and the total number of sensors detected by this mechanism is to be assumed as being the maximum number of sensors (or *True Positives*) present in the datasets.In the investigation to answer this research question, each snapshot within the datasets is evaluated using SensorBuster with varied *responsiveness* threshold R from 0.0 to 0.9.

Next, since the SensorRanker mechanism relies upon clustering algorithms to help distinguishing sensors from bots, the following research question also needs to be answered:

• *Which clustering algorithm is suitable in distinguishing sensors from bots when applied to the results of the SensorRanker mechanism?*

For that, the effectiveness of five clustering algorithms, namely *K-Means, DBSCAN, Gaussian Mixture Models, SpectralClustering*, and *Agglomerative Clustering* from the *scikit-learn* module is evaluated to be used in SensorRanker to classify sensors. These algorithms were chosen due to their simplicity (in operation) as they easily create two clusters from any given set of data, i.e., distinguishing between bots and sensors. The effectiveness of a clustering algorithm is evaluated by the highest number of classified sensors, i.e., *True Positives*, in combination with the lowest number of *False Positives*. This experiment is conducted with the *responsiveness* threshold of $R = 0.0$ for each snapshot within both datasets. Based on the results of the experiment, the best algorithm is then chosen and used for subsequent analysis.

As artifacts present in datasets can highly influence the accuracy of the detection mechanisms, the following research question needs to be answered in the evaluation:

- *How influential are the artifacts which are present in datasets towards the accuracy of the detection mechanisms?*

All three detection mechanisms are evaluated for each snapshot in both datasets with varying *responsiveness* threshold R from 0.0 to 0.9 to answer this question. It is expected that with a higher value of R, the number of artifacts or false positives decreases accordingly (if applicable).

It is also of interest to identify and evaluate the strengths of each of the proposed detection mechanisms. Therefore, the following research question needs to be answered in the evaluation:

- *Which mechanism performs best amongst the three sensor detection mechanisms?*

For this, the performance of all three mechanisms is evaluated and compared on both datasets. This evaluation is performed by selecting the appropriate clustering algorithm for SensorRanker and best values of R to eliminate the influence of artifacts within the datasets by answering the previous research questions.

Finally, the feasibility of the methods proposed to circumvent the detection mechanism in Sect. 5.2 need to be analyzed. Hence, the following research question needs to be answered in the evaluation:

- *How feasible are the proposed DSI strategies in circumventing the three detection mechanism?*

An analysis is conducted on the Sality dataset to demonstrate the feasibility of the proposed strategy to circumvent the three mechanisms.

5.3.4 Results

In answer to the research questions presented in Sect. 5.3.3, the evaluation results are presented.

Establishing Baseline Information

The SensorBuster mechanism is evaluated using both datasets with varying values of *responsiveness* threshold R 0.0–0.9 to establish the baseline information on the total available sensors. Analysis of the Sality dataset indicated a combined total of 61 nodes were flagged by the SensorBuster mechanism with the threshold value of $R = 0.0.11$ nodes were verified to be *Sensor* and seven of them were classified as *Unknown* based on manual analysis using the auxiliary data (see Sect. 5.3.2.2). However, all *Unknown* nodes were later found out to be artifacts. The remaining 43 nodes were false positives, i.e., churn affected nodes that were present for only a short period. Detailed results of the total sensors present in the different snapshots, i.e., days, are provided in Table 5.3 in dependence on the various values of R. A value within the table should be interpreted as the total number of sensors that were present on a particular day with a responsiveness ratio of at least R.

The analysis on the ZeroAccess dataset indicated a total of four nodes were flagged by the SensorBuster mechanism with the minimum value of $R = 0.0$. Out of these nodes, three nodes were verified to be *Sensor* and one was a false positive. These nodes were also consistently seen across all thresholds throughout the week.

Both sets of information for Sality from Table 5.3 and for ZeroAccess are then assumed as the *ground truth*. This data is then used for evaluation for the remainder part of this chapter.

Suitable Clustering Algorithm for SensorRanker

To investigate the effectiveness of different clustering algorithms in classifying sensors based on the *SensorRank* values, the evaluation was carried out on both datasets with the *responsiveness* threshold set to $R \geq 0.0$. In the results, the performance of *K-Means*, *Gaussian Mixture Models*, and *SpectralClustering* were identical across the different days and thresholds regarding the number of detected sensors and false positives. Therefore, these algorithms are grouped and referred to as *All Others* for clarity purposes. Figure 5.4 presents the performance of *DBSCAN*, *Agglomerative Clustering*, and *All Others* depending on the day of the measurement within the Sality dataset. The individual markers represent the number of detected sensors, and the bar plots indicate the number of false positives incurred by the algorithm for a given day. The results indicate that all algorithms except *DBSCAN* incurred exactly one false negative. *DBSCAN* incurred two false positives throughout the whole week. Also, there were no false positives generated by all detection mechanisms between day three and six (cf. Table 5.4).

Upon further investigation, it is discovered that all algorithms missed a particular node that had low popularity. This node which had only one incoming connection,

Table 5.3 Maximum sensors present on a particular day dependent on R in the Sality dataset

Day	Minimum responsiveness threshold, $R\geq$									
	0.0	0.1	0.2	0.3	0.4	0.5	0.6	0.7	0.8	0.9
1	10	10	10	9	9	9	8	8	7	7
2	11	11	11	11	10	10	9	9	8	8
3	11	11	11	11	11	11	11	11	10	9
4	11	11	11	11	11	11	11	11	11	10
5	10	10	10	10	10	10	10	10	10	9
6	10	10	10	10	10	10	10	10	10	10
7	10	10	10	10	10	10	10	10	10	10

Table 5.4 Effectiveness of different clustering algorithms with $R = 0.0$ on accurately classifying sensors in the Sality dataset

Day	Agglomerative		DBSCAN		All others		Total sensors
	TP	FP	TP	FP	TP	FP	
1	9	0	9	2	9	0	10
2	10	1	10	1	9	1	11
3	10	0	10	0	10	0	11
4	10	0	10	0	10	0	11
5	9	0	9	0	9	0	10
6	9	0	9	0	9	0	10
7	9	0	9	0	9	0	10

was an instance of BT node that was deployed within Sality (see. Sect. 4.2.2). However, since the nature of this BT node satisfy all characteristics of a sensor node (see Sect. 5.1.1.3), this node should have also been detected as a sensor. *DBSCAN* is observed to perform inferior to the other algorithms as it generated lesser false positives compared to the others.

The evaluation is repeated on the ZeroAccess dataset and all algorithm were found to successfully detect three sensors throughout the week with no incurred false positives or false negatives. After considering the analysis results of both datasets, the *Agglomerative Clustering* algorithm was picked to be used as the choice of clustering algorithm in SensorRanker for the rest of the analysis, including on the ZeroAccess dataset, due to its improved performance particularly on Day 2 within the Sality dataset.

Influence of Artifacts Present Within Datasets

As artifacts could adversely affect the performance of the detection mechanisms, their influence was investigated with varying values of R, i.e., between 0.0 and 0.9, on all three detection mechanisms using both datasets. Results indicate that high number of false positives were observed for both SensorBuster and LCC in the Sality dataset when $R = 0.0$. Both mechanisms managed to detect all existing sensors, i.e., 11, at the expense of 43 and 70 false positives respectively. Since the high number of false positives distort the overall representation of the results, the analysis of the Sality dataset is presented in Fig. 5.5 only for values of $R = [0.1, 0.9]$. This figure represents the number of nodes classified as *Sensor* by the respective detection mechanisms along with the corresponding false positives dependent on varying values of R.

The results of this analysis met the initial expectation presented in Sect. 5.3.3 that with increasing values of R, false positives can be reduced considerably. The observation of reduced false positives is particularly true for LCC that was able to reduce 90% of its false positives from the initial 70 to only seven false positives when R is set from 0.0 to 0.5. Although, similar observations were seen for SensorBuster, this was only observed for values of $R \leq 0.5$. Interestingly, SensorRanker was found to be least affected by the different values of R. It incurred only one false positive

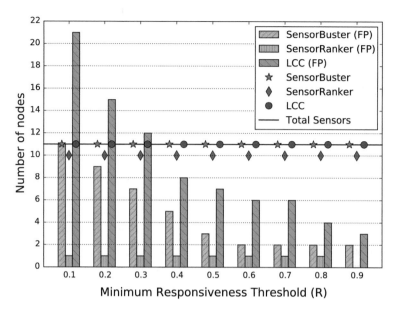

Fig. 5.5 Analysis of the influence of artifacts to the detection mechanisms with $R = [0.0, 0.9]$ on the Sality dataset

for all values of $R < 0.9$. However, the mechanism also incurred one false negative throughout the different values of R.

Similar observations were also seen in the evaluation for the ZeroAccess dataset. All detection mechanism were able to detect all present sensors, i.e., three sensors. However, SensorRanker did not generate any false positives regardless of the different values of R. Meanwhile, both SensorBuster and LCC eliminated their only false positive with threshold values of $R \geq 0.1$ and $R \geq 0.2$ respectively.

In conclusion, as presented in Table 5.5, only SensorRanker was found to be minimally affected (if any) by the different values of R. Hence, they would perform the same regardless of the presence of artifacts in the snapshot. Furthermore, it is also observed that LCC is heavily influenced by the presence of artifacts compared to the SensorBuster mechanism. As such, a minimum responsiveness threshold of $R \geq 0.6$ is found to be a conservative value for all detection mechanisms without accidentally ignoring sensors with poor responsiveness within the Sality dataset. Similarly, a threshold of $R \geq 0.2$ is observed to be appropriate for all detection mechanisms on the ZeroAccess dataset. The disparity between the threshold values of the Sality and ZeroAccess dataset can be argued as being directly influenced by the MM interval of the respective botnets. Sality, which has a longer MM interval compared to ZeroAccess, has a higher probability of introducing artifacts from stale nodes in their *NL*s. In contrast, the shorter MM interval of ZeroAccess reduces the probability for its bots to have stale entries in their *NL*s.

Table 5.5 Performance comparison of all three sensor detection mechanisms

	LCC	SensorRanker	SensorBuster
True positives	High	Medium	High
False positives	High	Low	Low
Sensitivity to artifacts	High	Low	Medium

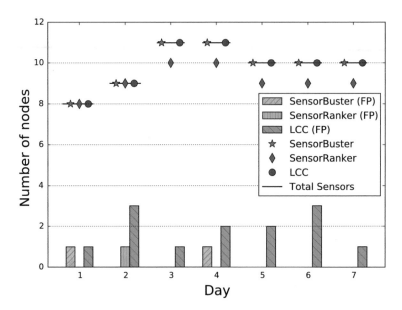

Fig. 5.6 Performance comparison of all detection mechanism with $R = 0.6$ on the Sality dataset

Performance Comparison of all Detection Mechanism

A comparative analysis of the performance of all three detection mechanisms is performed on both datasets with responsiveness threshold of $R = 0.6$ and $R = 0.2$ respectively. Figure 5.6 presents the results on the Sality dataset whereby the accuracy of each detection mechanism in dependence on the day of the snapshot is plotted. Meanwhile, Table 5.5 provides a brief summary of the performance comparison among the three detection mechanisms.

Figure 5.6 indicates that both SensorRanker and LCC were able to detect all sensors on each day. SensorRanker, although incurred one false negative compared to the other mechanisms, has the least number of false positive throughout the whole week. In comparison, LCC performs worst compared to all other mechanisms in terms of the number of incurred false positives. Meanwhile, the analysis of the ZeroAccess dataset indicated that all algorithms were able to detect exactly three sensors with no false positives or false negatives.

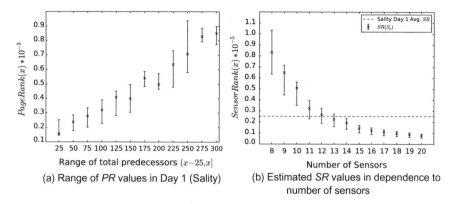

Fig. 5.7 Feasibility analysis of evading SensorRanker

Feasibility of Circumventing LCC

As discussed in Sect. 5.2.1, LCC can be circumvented using the DSI strategy with a minimum of four colluding sensor nodes. In fact, as long as the DSI strategy is able to obtain any non-extreme value, i.e., $lcc^+ \notin \{0.0, 1.0\}$, LCC can be circumvented. However, if the LCC-values of the sensors are significantly different than most of the bots in the botnet, clustering algorithms can be used to identify such anomalies in the values as used in the SensorRanker detection mechanism. Therefore, additional effort may be required to first identify the average value of the botnet, and the sensors should try to approximate this value using the DSI strategy to avoid being detected through the usage of clustering algorithms.

Feasibility of Evading SensorRanker

An analysis (cf. Fig. 5.7) was conducted in an attempt to provide a lower bound estimation of the number of sensors needed to evade the SensorRanker detection based on the Sality dataset. For this, sensors are assumed to evade detection if their *SensorRank* value is lower than or equal to the average value of all bots. As such, the minimum number of sensors $N_{min} = |S|$ that are required to satisfy Eq. (5.3) needed to be calculated to evade the mechanism.

However, it is not easy to calculate the PageRank (PR) values of a sensor S_i, as it is not only influenced by its predecessors but also by the rank of all of its predecessors. As such, the distribution of existing PR values for bots on Day 1 in the Sality dataset is referred. Figure 5.7a represents the *maximum*, *minimum* and *average* PR values with respect to the range of total predecessors in the investigated snapshot. The classification results indicate a linear increase in PR-values relative to increasing range of total predecessors, i.e., higher popularity yields higher ranks. Moreover, some ranges are also observed to have extreme PR values compared to their next ranges, e.g., compare the range of $(225, 250]$ and $(250, 275]$. This behavior is due to one or more predecessors of some bots within the range $(225, 250]$ having abnormally high rank, hence increasing the max PR of the bot(s) within this range compared to those in range $(250, 275]$.

Based on the distribution of PR values in Fig. 5.7a, all possible SensorRank (SR) values is calculated by considering scenarios of deploying up to 20 colluding sensors within the Sality botnet as presented in Fig. 5.7b with respect to the number of colluding sensors. The values plotted in this figure are the corresponding *maximum*, *minimum* and *average* SR values considering the possible range of total predecessors each of the sensors would have. As the average SR value for the whole dataset is 2.5145×10^{-6}, the maximum SR value of each colluding sensor needs to be less than it. Therefore, at least 14 colluding sensors are needed to safely evade the SensorRanker detection mechanism based on the utilized snapshot.

However, the estimation shown in Fig. 5.7b does not represent the worst case scenario, where one of the sensors has the $\frac{|V|}{N}$ highest ranked nodes as predecessors. Hence, the estimation provides only a best-case scenario, but in reality requires additional efforts to distribute the PR values intelligently or with more sensors to evade this detection mechanism. In conclusion, SensorRanker can be circumvented provided sufficient resources be available at disposal to deploy additional sensors.

Feasibility of Circumventing SensorBuster

As discussed in Sect. 5.2.3, to the best of knowledge, SensorBuster can only be circumvented if sensor nodes return valid bots when being requested. Instead of returning any bot in the NL replies, a *less useful* bot is suggested to be returned, i.e., newly joined bots. From the perspective of a sensor, this strategy incurs only minimal overhead to keep track of the timestamps of bots recently joined and known to the sensor. However, considering the fact that newly joined bots have the tendency to leave the botnet overlay immediately, the returned bot may not be responsive, i.e., offline, when a crawler captures the snapshot prior to the SensorBuster analysis. As a consequence, the sensor can once again be detected by SensorBuster due to the missing path(s) into the main SCC of the botnet. Therefore, it is advisable to return more than one of such newly joined bots, i.e., redundancies, to ensure there is at least one path into the main SCC.

5.4 Summary

This chapter focused on advanced monitoring on the basis of using sensor nodes in P2P botnets and outlined another major contribution as part of this book. In particular, contrary to the reports of other researchers, works presented in Sect. 5.1 indicates that it is indeed possible to detect deployed sensor nodes in P2P botnets. Evaluation results of the LCC, SensorRanker, and SensorBuster suggest that many existing sensors are currently susceptible to the proposed detection mechanisms.

Detecting Sensor Nodes

All proposed mechanisms were successful in detecting sensors deployed in Sality and ZeroAccess. However, LCC is more prone to false positives resulting from artifacts present in the datasets than SensorBuster. Meanwhile, SensorRanker is able to perform reasonably well in detecting sensors and only minimally influenced by artifacts. Therefore, a future botmaster would potentially deploy the SensorBuster mechanism if he is only concerned on detecting all sensors that are present in the botnet but tolerant to some false positives. However, this mechanism requires additional *responsiveness* information of the bots in addition to the connectivity information. In the absence of such information or if accurately identifying sensors is the only concern, the botmaster should utilize SensorRanker instead.

Circumventing the Detection Mechanisms

This chapter also introduced methods to circumvent the proposed detection mechanisms using a set of colluding sensors, i.e., DSI. While it is relatively easy to circumvent LCC and SensorRanker, it is more complicated to circumvent SensorBuster. As discussed in Sect. 5.2.3, it requires a sensor to return valid bots when being requested for neighbors to evade the detection of SensorBuster. As a result, such actions may have some legal implications as it could contradict with cyber-laws of many countries. In view of this, more work needs to be done to investigate the extent of which an organization or individual should be allowed to go in future botnet monitoring. This is especially important in anticipating future anti-monitoring countermeasures that could enforce strategies that require all bots, including sensors or crawlers, to participate in regular botnet maintenance activities before they can retrieve any information, e.g., additional neighbors. In the next chapter, this book is concluded and an outlook is presented.

Acknowledgements Parts of the contributions of this chapter is funded by Universiti Sains Malaysia (USM) through Short Term Research Grant, No: 304/PNAV/6313332.

References

1. Andriesse, D., Rossow, C., Bos, H.: Reliable recon in adversarial peer-to-peer botnets. In: ACM SIGCOMM Internet Measurement Conference (IMC) (2015)
2. Böck, L., Karuppayah, S., Grube, T., Mühlhäuser, M., Fischer, M.: Hide and seek: detecting sensors in P2P botnets. In: IEEE Conference on Communications and Network Security, pp. 731–732 (2015)
3. Hagberg, A.A., Schult, D.A., Swart, P.J.: Exploring network structure, dynamics, and function using NetworkX. In: Proceedings of the 7th Python in Science Conference (SciPy2008), vol. 836, pp. 11–15 (2008)
4. Page, L., Brin, S., Motwani, R., Winograd, T.: The PageRank Citation Ranking: Bringing Order to the Web. Technical report, Stanford InfoLab (1999)
5. Pedregosa, F., Varoquaux, G., Gramfort, A., Michel, V., Thirion, B., Grisel, O., Blondel, M., Prettenhofer, P., Weiss, R., Dubourg, V., Vanderplas, J., Passos, A., Cournapeau, D., Brucher, M., Perrot, M., Duchesnay, E.: Scikit-learn: machine learning in Python. J. Mach. Learn. Res. **12**, 2825–2830 (2011)

6. Stutzbach, D., Rejaie, R.: Understanding churn in peer-to-peer networks. In: Proceedings of the 6th ACM SIGCOMM Conference on Internet Measurement (2006)
7. Tarjan, R.: Depth-first search and linear graph algorithms. SIAM J. Comput. **1**(2), 146–160 (1972)
8. Watts, D.J., Strogatz, S.H.: Collective dynamics of "small-world" networks. Nature **393**, 440–442 (1998)
9. Yan, J., Ying, L., Yang, Y., Su, P., Li, Q., Kong, H., Feng, D.: Revisiting Node Injection of P2P Botnet. Lecture Notes in Computer Science, vol. 8792. Springer International Publishing, Cham (2014)

Chapter 6
Conclusion and Outlook

In the previous chapters, issues relevant to advanced P2P botnet monitoring were visited. Particularly, existing botnet monitoring mechanisms were analyzed regarding the challenges faced by them: the dynamic nature of P2P botnets and the various anti-monitoring mechanisms implemented by botnets. Based on that, this book proposed several countermeasures to circumvent existing anti-monitoring mechanisms. In addition, several new and advanced anti-monitoring mechanisms have also been introduced to anticipate the next steps of the botmasters. This chapter summarizes the main contributions and findings of this book and presents an outlook.

6.1 Conclusion

The adoption of a P2P-based architecture by recent botnets made monitoring them more difficult. Specialized monitoring mechanisms such as crawlers and sensors are required to monitor them. However, botmasters have equipped their botnets with anti-monitoring mechanisms that impede botnet monitoring. Examples of such botnets are GameOver Zeus, Sality, and ZeroAccess. To make things worst, the dynamic nature of P2P botnets also represents itself as a hurdle for botnet monitoring.

In this book, requirements for an advanced botnet monitoring mechanism was derived to serve as a guideline for a discussion of the current state of the art in Chap. 2. The proposed requirements are not only aimed at producing high-quality monitoring data, but also stealthier monitoring. Analysis revealed that many of the existing monitoring mechanisms only partially fulfil the non-functional requirements that were outlined in Sect. 2.1.2. Continued usage of such mechanisms may produce

S. Karuppayah, *Advanced Monitoring in P2P Botnets*, SpringerBriefs on Cyber Security Systems and Networks, https://doi.org/10.1007/978-981-10-9050-9_6

biased results, and introduce noise that adversely affect the monitoring results of others.

Moreover, anti-monitoring mechanisms of existing botnets also cause the existing monitoring mechanisms fail in fulfilling the Stealthiness or Accuracy requirements. One example of such a mechanism is the NL restriction and automated blacklisting mechanism of GameOver Zeus that targets crawlers. In this book, a countermeasure called ZEUSMILKER is proposed to circumvent this NL restriction mechanism of GameOver Zeus that provably retrieves all neighbors of a bot. Then, to address the issue of evading the automated blacklisting mechanism of GameOver Zeus, a novel crawling algorithm called (LICA) is proposed which minimizes the number of bots that needs to be crawled to enumerate bots in the botnet.

However, it is just a matter of time before the botmasters come up with newer mechanisms to impede monitoring. For this reason, this book also introduces several anti-monitoring countermeasures from the perspective of a botmaster to raise the stake in this arms race. In particular, a lightweight crawler detection mechanism called (BTs) is proposed that leverages design constraints of existing botnets. Evaluation results of these mechanisms on Sality and ZeroAccess dataset indicates that many crawlers are susceptible to them. This is indeed worrying, since the idea of BT is relatively simple and can be easily implemented in existing botnets.

Prior to the work done in this book, some researchers have claimed that sensors are a more stealthy monitoring mechanism as they are indistinguishable from bots. However, as another major contribution of this book, three sensor detection mechanisms were proposed by leveraging graph-theoretic metrics to discern sensors from bots. Evaluation results of these mechanisms in Sality and ZeroAccess indicates that many sensors are susceptible to the proposed mechanisms. Particularly, the *Sensor-Buster* mechanism can accurately detect independent and colluding sensors deployed in a botnet. To give the upper hand back to the defenders, this book also discussed strategies that should be adopted by future sensors to remain undetected from the proposed sensor detection mechanisms.

One major problem that can be foreseen with the advancement of botnet monitoring mechanisms is that most of these mechanisms would become stealthier than now. While being stealthy may help to perform monitoring, this may also introduce adverse effects, i.e., unnoticeable noise, to the monitoring data of others. For instance, the stealthy monitoring footprint of an individual or organization will be assumed by others to be those of the botnet itself. This in turn would affect the accuracy and quality of the monitoring data.

Concluding, the works presented in this book indicate that the existing anti-monitoring mechanisms implemented by botnets or proposed by researchers are still in their infancy. They can either be circumvented or tolerated with sufficient resources at disposal. However, botmasters are expected to improve and introduce more advanced anti-monitoring mechanisms. For that, the defenders need to be prepared to face such advancements. This can be done by preempting the possible advancements from botmasters, i.e., proposing advanced anti-monitoring mechanisms, and attempt to design countermeasures against them.

6.2 Outlook

This book primarily focused on proposing countermeasures to circumvent or tolerate existing anti-monitoring mechanisms. However, it would be interesting to identify to which extent botnet monitoring can always be performed. To answer this question, an in-depth investigation and analysis is required starting with the set of assumptions that are applied on both the botnets and the monitoring mechanisms themselves. For instance, crawling can enumerate bots by leveraging the MM protocols of the botnets. If future botnets does not require exchange of neighbors to maintain connectivity with the overlay, crawling may no longer be useful. Therefore, understanding the possible extents of monitoring will help defenders to prioritize and focus in developing new countermeasures or monitoring mechanisms.

The monitoring data collected in the context of this book also requires additional analysis. For instance, it is important to understand the impact of unknown third party monitoring activities on monitoring data, e.g., network properties like average path length. The need to return valid bots to circumvent the SensorBuster mechanism may be illegal from the perspective of cyber-laws of certain countries. Therefore, it may be necessary to revise the laws to grant some flexibility to the defenders when monitoring.

In addition, circumventing IP-based anti-monitoring mechanisms like those of GameOver Zeus requires a defender to possess a large pool of IP addresses at his disposal. Since this is not always feasible, there should be a focus in developing a collaborative botnet monitoring ecosystem that can utilize a large pool of shared IP addresses and resources from interested users and organizations, e.g., collaborative and distributed botnet crawling.

Finally, existing anti-monitoring mechanisms that were analyzed in this book can be either circumvented or tolerated with sufficient computing and network resources. Therefore, to anticipate further advancements from the botmasters, more work from the perspective of a botmaster needs to be carried out. Particularly, the author believes that future botnets may attempt to restrict further the botnet information that are shared among bots to impede botnet monitoring. However, this exchange of information among bots itself is important for the management of the botnet's overlay. Therefore, it is interesting to investigate the impact of such restriction mechanisms or botnet design to the robustness and resilience of the constructed botnet overlay.

Erratum to: Advanced Monitoring in P2P Botnets

Erratum to:
S. Karuppayah, *Advanced Monitoring in P2P Botnets*,
SpringerBriefs on Cyber Security Systems and Networks,
https://doi.org/10.1007/978-981-10-9050-9

The original version of the book was inadvertently published without incorporating the following corrections:

Some figures and tables excerpted from a published IEEE article and an ACM article should be included in Chapters 3, 4 and 5. The author has obtained permissions from both the articles but did not add credit lines about the copyright information in the book, which should to be now added.

In Chapter 5, acknowledgment should be newly included.

The book has been now updated with the changes.

The updated online version of these chapters can be found at
https://doi.org/10.1007/978-981-10-9050-9_3
https://doi.org/10.1007/978-981-10-9050-9_4
https://doi.org/10.1007/978-981-10-9050-9_5

Printed in the United States
By Bookmasters